What's in a Star?

STUDENT MANUAL

W9-AUC-195

What's in a Star?

STUDENT MANUAL

Melissa C. Kido
University of California, Berkeley, CA

Susan E. Kegley
University of California, Berkeley, CA

Greg Bothun
University of Oregon, Eugene, OR

Geoffrey W. Marcy
University of California, Berkeley, CA

W. W. NORTON & COMPANY
NEW YORK LONDON

Copyright © 2004 by the Trustees of Beloit College and the Regents of the University of California. Published through exclusive license by W. W. Norton & Company.

This module has been developed under the direction of the ChemLinks Coalition, headed by Beloit College, and the ModularChem Consortium, headed by the University of California at Berkeley. This material is based upon work supported by the National Science Foundation, grants No. DUE-9455918 and DUE-9455924. Any opinions, findings, and conclusions or recommendations expressed in this material are those of the authors and do not necessarily reflect the views of the National Science Foundation, Beloit College, or the Regents of the University of California.

This project was supported, in part by the Division of Undergraduate Education of the
National Science Foundation
Opinions expressed are those of the authors and not necessarily those of the Foundation

All rights reserved.

Reproduction or translation of any part of this work beyond that permitted by Section 107 and 108 of the 1976 United States Copyright Act without the permission of the copyright holder is unlawful. Requests for permission or further information should be addressed to the Permissions Department, W. W. Norton & Company

ISBN 0-393-92644-3

Acknowledgments
Many individuals have contributed valuable advice
and ideas. The authors would like to thank Tricia Fer-
ret, Marco Molinaro, Angy Stacy, Eileen Lewis, Jenni-
fer Loeser, George Lisensky, and Truman Schwartz
for their helpful input. (if cover image is used) The
cover picture is from the NASA collection of star pictures.

Contents

SESSION 4: WHAT DO STELLAR SPECTRA TELL US? (PART II)   47

Many-electron atoms

SESSION 5: WHAT IS IN A STAR?   74

Culminating Project

WHAT CAN YOU LEARN FROM THE MODULE?

This module challenges you to answer the question "What is in a star?" Over a period of 2-3 weeks, you will view images of stars and investigate the properties of stars and starlight. You will explore connections between the temperature and color of a star, and the radiation it emits. You will also closely examine the link between starlight and what this tells you about a star's structure and composition. We emphasize an understanding of atomic structure and processes such as absorption and emission of radiation by atoms by asking you to use this chemistry to answer questions about the nature of stars.

Interdisciplinary Aspects. As you learn about the chemistry of stars you will also learn about how it overlaps with other disciplines and areas of your life, including astronomy, physics, optics, and spectroscopy.

Level and Prerequisites. This module is designed to be covered in the first semester of a general chemistry course; no prerequisites are required.

Chemistry Content Objectives. This module will ask you to develop a working knowledge of the following:

∑ Electromagnetic spectrum: wavelength, frequency, and the different types of radiation.
∑ Properties of radiation: diffraction and interference, and the photoelectric effect.
∑ Blackbody radiation.
∑ Interaction of light with matter: emission, absorption, and transmission of light by objects, and visible spectroscopy.
∑ Atomic Structure: atomic spectroscopy and quantum mechanics.
∑ Periodic Trends: ionization energy, atomic radii and electron affinity.

Scientific Skill Objectives. This module will also ask you to develop the following skills.

∑ Data analysis and interpretation skills, quantitative and qualitative.
∑ The formulation and application of models, including comparisons to data and evaluation of models and predictions.
∑ Reasoning with a microscopic view (on the level of atoms and molecules) in order to explain and predict macroscopic (observable) properties.
∑ Reading and writing skills: using evidence to support or refute a hypothesis, explaining concepts, summarizing logic of a scientific argument.
∑ World Wide Web and computing skills.

HOW CAN YOU LEARN FROM THE MODULAR APPROACH?

This manual includes inquiry-based activities for the classroom, for the laboratory, for the computer laboratory, and for homework. What do we mean by inquiry-based activities? Instead of giving you the answers and requiring you to memorize fact, we give you questions that will lead you to a deeper understanding of chemistry concepts and approaches used by scientists in solving problems. The curriculum requires active participation from you, not only in the laboratories but also in the classroom. You will be recording observations from demonstration, doing experiments, solving problems, discussing results, constructing and evaluating models to explain your observations and analyses, and writing about what you have learned.

Organization of the Module

The module begins with a question that provides a context for understanding and exploring chemistry concepts.

Sessions. The module is divided into sessions, each beginning with a Session Question. Each session focuses on one aspect of the Module Question and will raise the issues you need to consider to respond to the Module Question.

Explorations. The Session Questions will be examined in classroom, laboratory, and computer Explorations. Each Exploration begins with a question that considers the Session Question in more depth and in different ways. *Your instructor will choose which Explorations to use* depending on her or his goals and the needs of your class. Each Exploration is divided into three or four sections:

1. **Creating the context.** This section states the goal of the Exploration and frames the Exploration Question. The text discusses why the question is important, reminds you of information discussed previously, gives background information, and describes what you can do to find out more. **Bold** type is used for key chemistry concepts.

2. **Preparing for Inquiry.** This section contains background reading, activities, or questions to help prepare you on your own for the main activities in the Exploration. In a laboratory exploration, this section would include pre-laboratory questions.

3. **Developing Ideas.** This section describes how to gather the information you need to respond to the Session and Exploration Questions. Activities of the following types are used for this guided inquiry.

    a. *Demonstrations* or data will be provided by your instructor or manual, or shown on a video or a computer. You'll need to understand the purpose of the data or experiment, make detailed observations, and attempt of explain these observations.

b. *Small group discussion and class discussion* will help you formulate responses to questions, gather data from other groups, and give you different perspectives.

c. In a *laboratory exploration*, you will design and carry out experiments to respond to a specific question. These experiments will require you to use chemistry principles and thinking skills related to you work in the classroom.

d. *Computer assignments* will ask you to use the computer to gather information, design experiments, and even run them.

4. **Applying Your Ideas.** The questions in this section are provided to help you explore the observations and data you have collected. Your instructor will choose which of these questions to use for interactive discussion, small group work or homework.

Looking Back. Each Session ends with a section called Looking Back. This Looking Back provides you with a list of the chemistry concepts you have learned, indicated by a **bold** type, and a list of scientific thinking and problem-solving skills that are appropriate to put on a resume.

Checking Your Progress. Each Session ends with a section called Checking Your Progress. This section provides you with activities that help you integrate and reflect on the ideas you have learned, connect your learning to the Session Question and storyline, and make progress toward the culminating activity.

Culminating Project. As the Module ends, you will be asked to do a project that will integrate everything you have learned.

General Approach

Learning in a Context. You may wonder about the necessity of a context for learning chemistry. You may feel that you could just learn the concepts and not bother learning the contextual information. Research in cognitive science has shown, however, that we retain knowledge better when appropriate background material is available. As an example, try the following experiment. Read the following paragraph once, close the book, and write down as much as you remember.

The procedure is not difficult. First, bring 1 liter of water to a state where it has undergone partially a phase transition in which the vapor pressure of the steam that is formed is equal to the pressure of the atmosphere. Then add 1.0 of the mixture of chemicals known as camilla thea. The important ingredient in this mixture is 3,7-dihydro-1,3,7-trimethyl-1H-purine-2,6-dione. Allow the mixture to stir for 5 minutes. Finally filter the undissolved solids and collect the liquid.

If you are unfamiliar with the technical language, you may not be able to recall much of these instructions. However, if you are told that the passage is about making tea, suddenly you can figure out much of the new vocabulary and enhance your retention of the instructions. The context has helped you use background knowledge to comprehend the passage. We have built a context surrounding the chemistry concepts and the problem-solving skills that we hope you will learn in this module. We hope that this context will help you make links to your background knowledge and to experiences from you everyday life. We believe that making these connections will aid your comprehension and enhance your retention.

Transforming Knowledge. As part of this module, you will also be encouraged to find solutions to problems that cannot be solved simply by "telling" knowledge. At this stage in your academic careers, you need to make the transition from simply "telling" facts you have memorized to "transforming" your knowledge to solve new problems. This is what scientists and professionals such as doctors do. As an example, consider the type of problem that a medical doctor must solve.

A patient who is quite ill is examined by Dr. Rosario. After examining the patient the doctor realizes that he has never encountered this set of symptoms before, or at least he doesn't remember having encountered them. Imagine Dr. Rosario giving one of the following responses to the patient:
1. "Sorry, I can't help you. I can't find the answer in my text book."
2. "Sorry, I can't help you. There is something wrong with your blood chemistry, but I can't remember what I learned in first-year chemistry.
3. "Sorry, I can't help you. I wasn't told about symptoms like yours in medical school."

Clearly, doctors or scientists cannot be familiar with every case or remember all the information they learned in their training. But if they have learned how to approach complex problems, they will be equipped to deal with new issues. Therefore, they must know how to approach complex problems.

In this module, we hope to give you experience with approaches used by scientists to solve problems, especially those for which an answer is not immediately obvious and for which multiple solutions are possible. You will find that real-life problem solving is an iterative process, When you do not know how to solve a problem, you start by exploring your best ideas. If these do not lead you toward a solution, you may have to back-track, rethink your ideas, and try something else. This process of generating and then refining your ideas allows you to define the problem more clearly. Eventually, you may reach an acceptable solution. As you generate and refine ideas about the issues and concepts in this module, your thinking will become more sophisticated, and your understanding will deepen.

Enjoy!

Melissa C. Kido
Susan E. Kegley
Greg Bothun
Geoffrey W. Marcy

What is starlight?

The Nature of Light

Exploration 1A

What are your initial ideas about stars?

Creating the Context

For eons, humans have looked up at the stars in awe and wonder, marveling at their beauty and their constancy. Scientists gaze at the stars in hope of gaining clues to the origins of the universe and to see how our star, the Sun, fits into the grand scheme of the universe.

The Sun is but one of a vast number of stars, about two hundred thousand million grouped together in a disk known as the Milky Way system or galaxy. Many of these stars populate our night sky. Many millions more are too faint to be detected with the naked eye. Our galaxy in turn is one of thousands of millions of systems that lie in the observable universe. (To see a Hubble telescope deep field image go to: http://www.phy.mtu.edu/apod/ap980607.html)

As the source of energy for all life on earth, the Sun is critical to our survival. Its light energy powers the engine of plant photosynthesis, which in turn supports the remainder of life on earth. Its heat energy warms our planet to a habitable temperature.

With the recent discoveries of at least 17 stars that, like our Sun, have one or more planets, people now are asking: Are other earth-like planets out there? Are there other life forms elsewhere in the universe? How can we learn more about these other places in the universe where life might arise? We seek answers to these questions by finding out more about the stars that form the centers of newly discovered planetary systems.

In this module, we will learn more about stars—what they are made of, what their temperatures are, and how they might be different from or similar to our Sun. A knowledge of chemistry is essential to understanding these aspects of stars. We begin by learning about the nature of light and energy, proceed to examine the relationship of temperature to a star's energy output, and finish by examining the structure of the atoms and molecules that comprise stars and their atmospheres.

Developing Ideas

In this module we will be developing a model of a star that describes its composition, structure, temperature, and color, but before we begin, take some time to think about what you already know about stars.

1. Work with your neighbor to list several things you currently know about stars.

2. Observe several images of stars and stellar objects. Record your observations of each object. These images can be found on the ChemConnections CD-ROM and web site in Exploration 1A. Alternatively, your instructor will provide the internet address for this page to you.

Astronomical Object	Observations

3. In small groups or with a neighbor, discuss and write down 2 or 3 questions you have about stars.

Applying Your Ideas

4. "We understand the possibility of determining their shapes, their sizes and motion, whereas never, by any means, will we be able to study their chemical composition."-*Auguste Compte, Cours de Philosophie Positive (1835)*

    The French philosopher Auguste Compte concluded in the early 19th century that it was impossible to know the composition of stars, since no one could bring a piece of a star into the laboratory. Yet, as we enter the 21st century, astronomers do have information about the composition of stars. Write down one or two ideas about how we might be able to obtain information about stars without actually going to them.

Exploration 1B

What are the characteristics of starlight?

Creating the Context

The most readily observable property of a star is the light it gives off. You might have noticed in the pictures of astronomical objects you observed in Exploration 1A that they have different colors and varying brightness. These properties of stars provide clues about their composition and temperature. Thus, if we are to know more about what is in a star, it is essential that we know more about starlight. In this Exploration, we will begin by observing several measurable properties of light and answer the question, *What are the characteristics of starlight?*

Preparing for Inquiry

One way we detect energy from stars is through the white light and colors we see in starlight. However, visible light constitutes only a small fraction of the total energy emitted by a star. Lower-energy radio waves and heat are also emitted, along with higher-energy X-rays. Although the energy of light is variable, different kinds of light have a number of characteristics in common.

The word *light* is the common name that we give to a form of energy called **electromagnetic radiation**. This technical name refers to the electric and magnetic fields associated with light energy. The electric and magnetic fields are vibrating and can be described as waves that move up and down in space and time (see Figure 1-1 and Figure 1-2).

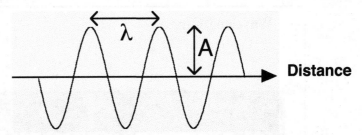

Figure 1-1: The graph of a wave versus distance at some instant in time. A is the amplitude and λ is the wavelength.

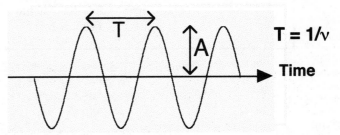

Figure 1-2: The graph of a wave vibrating in time. T is the period and ν is the frequency.

Figure 1-1 shows how an electric or magnetic field forms waves as it vibrates up and down periodically *with distance*, at a given instant in time. To see such waves on a vibrating string for example, we can take a snapshot of the string. The high points of the waves are called **crests** and the low points are called **troughs**. The distance between successive identical parts of the wave (from the top of one crest to the top of the next crest) is called the **wavelength**, represented by the Greek letter *lambda*, λ. The **amplitude**, A, of the wave is the distance from the vertical midpoint of the wave to the crest and is related to the **intensity** (brightness) of the radiation, the amount of light that strikes a unit area.

Figure 1-2 shows how an electric or magnetic field forms waves as it vibrates up and down periodically *in time*. The horizontal axis is now in units of time, and a new quantity, called the **period** T, is defined as the time required for one complete oscillation (from one crest or trough to the next crest or trough) of the wave. The **frequency** of the wave is the number of wave crests (or troughs) that travel past a certain point in a specified time. Frequency is symbolized by the Greek letter *nu* (represented by ν) and is related to the period by $\nu = 1/T$.

Frequency is also related to wavelength by the **speed of light**, c, which is 3.00×10^8 meters per second (m/s), as:

$$\nu = c/\lambda.$$

Because of the inverse relationship between frequency and wavelength, short wavelengths correspond to high frequencies and long wavelengths correspond to low frequencies. As the name implies, wavelengths are reported in units of length such as meters (m), centimeters (1 cm = 1×10^{-2} m), micrometers ($1\mu m = 1\times10^{-6}$ m), or nanometers ($1nm = 1\times10^{-9}$ m). Frequency has units of cycles per second or, more commonly, 1/s or s^{-1}. The unit s^{-1} is also called a **Hertz** (Hz).

The entire range of light or electromagnetic radiation is called the **electromagnetic spectrum** (See Figure 1-3). Although radio waves, microwaves, infrared radiation, visible light, ultraviolet radiation, X-rays, and gamma (γ) rays may seem at first consideration to be vastly different, they all are electromagnetic radiation. All of these forms of electromagnetic radiation carry energy, from the high-energy **ultraviolet (UV)** radiation that causes sunburn to the low-energy **infrared (IR)** radiation that warms us when we stand in front of a glowing heater coil.

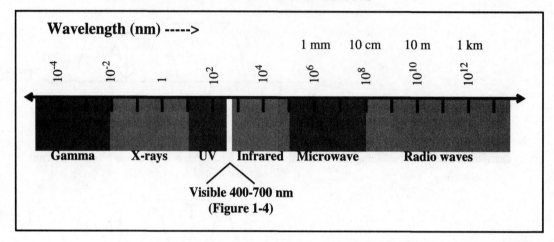

Figure 1-3:   The electromagnetic radiation spectrum spans a large range of wavelengths of light energy. The figure depicts the energies of the different forms of electromagnetic radiation in order of increasing wavelength from left to right.

The colors we enjoy as humans are part of the **visible** region of the electromagnetic spectrum, where light has wavelengths of about 400-700 nm (a nanometer (nm) is 10^{-9} m or about one trillionth of a meter long). The visible range is only a small fraction of the possible wavelengths of electromagnetic radiation and extends from violet light (high energy) to red light (low energy). White light can be separated into all the colors of the visible spectrum (see Figure 1-4). Next to the lower-energy red part of the spectrum is the infrared (IR) and next to the high-energy violet part of the spectrum is the ultraviolet (UV). Both IR and UV light are invisible to our eyes.

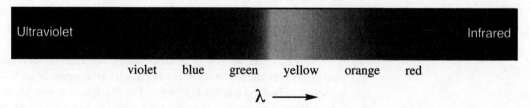

Figure 1-4:   Visible spectrum, with wavelength increasing from left to right.

Radiation can have wavelengths as short as 10^{-15} m for gamma rays, and as long as 10^3 meters (about 3/5 of a mile) for long-wave radio and 10 km for military submarine communication systems. Other wavelengths longer than visible light include infrared radiation and microwaves, while regions of shorter wavelengths include ultraviolet radiation and X-rays.

In this Exploration, you will investigate the different types of electromagnetic radiation, and the relationship between the wavelength and the frequency of electromagnetic waves.

Developing Ideas

1. For the following questions you will use the **Explore the Electromagnetic Spectrum tool** found on the Module Web Tools CD-ROM in Exploration 1B and on the web site. Alternatively, your instructor will provide the internet address for this page to you.

   a. Select *two* of the types of radiation in Table 1-1 below. Compare these wavelengths of light with the size of objects in the world around you. List one object that is of similar size to each wavelength of light.

   b. Using the relationship $\nu = c/\lambda$, calculate a representative frequency for each type of radiation, given the wavelengths in Table 1-1.

Table 1-1

EM Radiation	Example Astronomical Object	Wavelength (m)	Frequency (s^{-1})
X-ray	Supernovae	10^{-10}	
Ultraviolet (UV)	Hot stars	10^{-8}	
Visible (green light)	Sun	6×10^{-7}	
Infrared (IR)	Comet	10^{-5}	
Microwave	Interstellar gas	10^{-3}	
Radio	Solar flare	10	

Applying Your Ideas

The following exercises will help you better understand the different types of electromagnetic radiation and the relationship between wavelength and frequency. You may need to refer to your textbook for unit conversions.

2. The strongest extra-terrestrial source of radio waves we experience on earth is the Sun. Solar flares emit bursts of radio waves having wavelengths from 1 to 60 m. Express these wavelengths as frequencies (Hz or s^{-1}). Compare these frequencies to a typical FM radio broadcast frequency of 88.5 MHz (1 MHz = 10^6 Hz). Would you expect solar flares to interfere with your radio broadcast? (Additional solar flare information is at http://hesperia.gsfc.nasa.gov/sftheory/flare.htm)

3. As Comet Hale-Bopp approaches the Sun, it is heated and begins to eject tiny dust grains. Astronomers "see" comet dust because of the radiation, between 8 and 13 μm (micrometers), it gives off. In what region of the EM spectrum does this radiation belong? How would your body detect this radiation (as color, heat, a sunburn)? (Additional comet information is at http://seds.lpl.arizona.edu/billa/tnp/comets.html)

4. Supernovae such as Eta Carinae (see "Star images" in Exploration 1B on the ChemConnections CD-ROM and web site) are violent explosions of stars so bright that they can be seen even at great distances with the naked eye. In addition to this visible light, large quantities of X-rays are also emitted. A typical X-ray frequency is 3×10^{18} Hz. How does this wavelength compare to the wavelength of radiation emitted by comet dust in Problem 3 above?

5. By looking at radiation wavelengths between 1.3 and 2.6 mm, astronomers have found evidence of carbon monoxide (CO) in the interstellar medium between the stars. In what region of the electromagnetic spectrum does this radiation belong? Name one common use of this type of radiation.

6. Another common unit of length is the Angstrom, Å, where $1 \text{ Å} = 10^{-10}\text{m}$. Measurements of radiation at 6563 Å indicate that there is hydrogen in the interstellar medium as well as carbon monoxide. What is this wavelength in nanometers (nm)? Would you be able to see this radiation with your eyes?

Exploration 1C

How does light behave as a wave?

Creating the Context

In Exploration 1B you learned that light can be described as electromagnetic waves that have properties of wavelength and frequency. In addition to these properties, waves will bend around corners and will simultaneously pass through two slits or small openings in a barrier. In this Exploration we will consider these two properties of waves and begin to deepen our understanding of the nature of light by answering the question, *How does light behave as a wave?*

As an example of wave behavior you are likely to be familiar with, let's consider the interaction of water waves with solid objects. When water waves encounter an edge or a corner such as a rock or a concrete pier, the wavefront will appear to bend around the corner and spread out as part of the wavefront is cut off by the obstacle. This wave behavior is called **diffraction.**

Light will also diffract when it encounters an edge or corner. Figure 1-5 demonstrates the diffraction of light through a single slit. The arrows or **rays,** represent the direction of the traveling waves. The lines or **wavefronts,** show the location of the crests of the wave. Before the opening, the wave moves in one direction, and the wavefronts are straight lines. After part of the wave passes through the barrier, wavefronts proceed straight ahead *and* to the sides. If the light is projected on a screen far away, a distinct **diffraction pattern** is produced. The alternating light and dark bands are called *fringes* where light and dark fringes correspond to maximum and minimum light intensity, respectively.

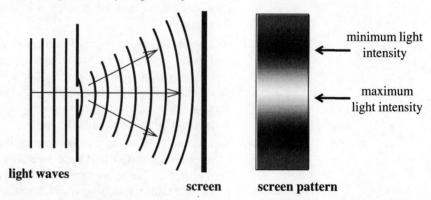

light waves   screen   screen pattern

minimum light intensity

maximum light intensity

Figure 1-5: The diffraction of light through a single slit. The wavefronts (lines) spread out or bend after passing through the barrier. The rays (arrows) show the direction of travel of the wave. The pattern produced on the screen is called a diffraction pattern.

When two waves arrive at the same region of space, **interference** can occur. Figure 1-6 shows examples of interference between two waves. **Constructive interference** (A) occurs when the crests (or troughs) of waves 1 and 2 coincide, producing a wave with *increased* amplitude. However, when the crests (or troughs) of waves 1 and 2 do not coincide, **destructive interference** (B) produces a wave with *decreased* amplitude. Figure 1-6 (B) shows an example of destructive interference where the crests of one wave coincide with the troughs of the other, resulting in a wave with zero amplitude. One way to cause interference is to pass a wave through a barrier with two openings in it. The two waves that emerge from the two slits will diffract and overlap, or interfere, with each other.

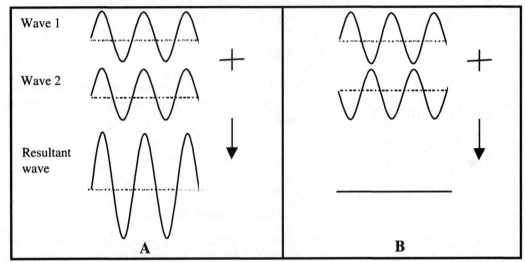

Figure 1-6: Constructive interference (A) and destructive interference (B) of waves 1 and 2.

Developing Ideas

Diffraction of Light Through a Single Slit

1. Consider light as a particle.

    a. How will a stream of light particles appear after passing through a single slit in a barrier shown below? Sketch the *stream of particles after slit* in the space provided below.

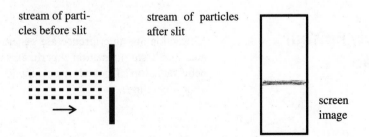

    b. If you projected the resultant stream of particles of light onto a screen, what would you see? Sketch the *screen image* in the space provided above.

    c. Predict and sketch below how the screen image will change if you *increase* the width of the slit.

2. **Diffraction of Light through a Single Slit.** Your instructor will demonstrate the diffraction of light from a laser through a single slit. Observe how the image produced on a screen changes as the slit width increases.

   a. Sketch below the initial screen image produced by diffraction from a very narrow slit.

   b. Now, your instructor will slowly *increase* the width of the slit. Describe in words what you observe.

   c. Does the effect of increasing the slit width that you have observed agree with your prediction in Problem 1b for the case when light is a particle? Explain.

Diffraction of Light Through Two Slits

3. Your instructor will demonstrate what happens when light passes through two slits in a barrier. Observe and sketch the pattern produced on a screen in the space below.

4. Briefly describe the two-slit diffraction pattern that you observe. How does this pattern compare to the single-slit diffraction pattern from Problem 2?

Applying Your Ideas

5. Diffraction and interference are general wave phenomena. What do the single and double slit diffraction experiments indicate about the nature of electromagnetic radiation? Does electromagnetic radiation behave like a wave? Explain your reasoning.

Exploration 1D

How does light behave as a particle?

Creating the Context

Thus far in this Session, we have described light as a wave. In Exploration 1C you saw that diffraction and interference patterns are produced when light passes though one or more small openings in a barrier. That particular experiment supports a wave theory of light. Can we be sure, however, that light doesn't have particle-like characteristics?

By the end of the nineteenth century, most scientists supported a wave theory of light. According to this theory, the energy of light is determined by the amplitude or intensity of electromagnetic waves. However, in 1900, a German physicist named Max Planck proposed the existence of "particles" or packets of radiation, later called **photons**, in which the energy of a packet of radiation is proportional to frequency. Many scientists found it difficult to believe a particle theory of light; they needed experimental evidence to support the theory. In this Exploration, we will examine one such experiment and answer the question, *How does light behave as a particle?*

Astronomers who are interested in analyzing starlight often use instruments called **detectors**. We can think of a detector as any device in which some measurable property changes in response to incident electromagnetic radiation. One common example is sunglasses that darken in response to sunlight. Historically, astronomers used the human eye and photographic plates to detect light gathered by a telescope. These detectors now have largely been replaced by electronic devices such as CCDs (charge coupled devices) in large telescopes. They are often made of silicon, but other elements and mixtures of elements can be used (for an example of a CCD, see http://ccd.ifa.hawaii.edu/images/lab.gif). In order to understand how detectors measure the light from stars, we need to learn more about the nature of light, and whether it behaves as a particle, a wave, or both.

Preparing for Inquiry

The Photoelectric Effect: Using Light to Remove Electrons

Some modern detectors such as CCDs rely upon the discovery, made in 1887 by Heinrich Hertz, of the photoelectric effect. The results of this experiment greatly puzzled scientists until they were explained by Albert Einstein in 1905.

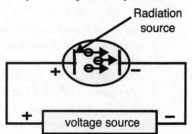

Figure 1-7:   Current flows in a photocell when radiation strikes a metal surface, causing electrons to be ejected from the metal and into the circuit.

Hertz demonstrated that shining radiation on metals can generate an electric current. With the apparatus shown in Figure 1-7, referred to as a photocell, Hertz observed that when radiation shines on a metal plate in a circuit, the radiant energy is converted into the energy of motion, or **kinetic energy**, of electrons. The electrons are ejected from the metal with a variety of kinetic energies and move toward another metal plate. The flow of electrons in this **photoelectric effect** is measured as **current** in units called Amperes (A). The amount of current measured is proportional to the number of electrons produced by the radiation.

More About Atoms

The photocell works because light energy causes electrons to be emitted from the metal surface, creating a current proportional to the intensity of the light. Similarly, the detection of light by CCDs relies on the flow of electrons. To begin to understand how radiation interacts with electrons in the photoelectric effect, we need to know more about the structure of an atom.

Electrons are negatively charged particles that are part of atoms. Electrons are very light particles (see sidebar), held in atoms by an attractive force exerted by the more massive positively charged **protons,** particles in the **nucleus** or center of the atom (see Figure 1-8). Energy is required to overcome this attractive force between an electron and the nucleus; the photoelectric effect, for example, uses light energy to eject electrons.

The nucleus also contains **neutrons**, particles of approximately the same mass as the proton but with zero charge. Elements, substances such as gold, helium, and oxygen, are distinguished from each other by the number of protons they have in the nucleus. The elements are listed on the periodic table in order of increasing number of protons, referred to as their **atomic number** and abbreviated **Z**. The **mass number**, **A**, is the sum of protons and neutrons for a particular element.

Masses of Subatomic Particles	
electron	9.1095×10^{-31} kg
proton	1.6725×10^{-28} kg
neutron	1.6750×10^{-28} kg

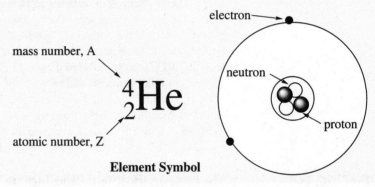

Element Symbol

Figure 1-8:   A simple model of the helium atom showing the nucleus containing protons and neutrons, with electrons in a shell outside the nucleus. The element symbol provides information about the number of protons (the atomic number Z) and neutrons in the element.

While all atoms of a particular element contain the same number of protons, they do not necessarily contain the same number of neutrons. Elements with the same number of protons but different numbers of neutrons are called isotopes of the element. There is a natural abundance of the stable forms of the different isotopes, with one isotope usually dominating. Many other unstable isotopes can exist for short periods of time before they undergo radioactive decay. Because protons and electrons, not neutrons, are largely responsible for the chemical reactivity and physical properties of elements, different isotopes of the same element are almost identical in their physical and chemical properties.

Developing Ideas

Results from the photoelectric effect experiment are illustrated in the diagram below. In this experiment, when light of a given frequency and intensity shines on a metal surface, electrons may be emitted. The ejected electrons are measured as current. Work in small groups to answer the following questions.

1. According to the **wave model** of light, the energy of radiation depends only on its intensity (wave amplitude), not frequency. However, the **particle model** of light says that the energy of radiation depends only on its frequency, not its intensity. (Instead, under the particle model, intensity corresponds to number of light particles.) These two models predict different results for the photoelectric effect experiment.

   a. Given a *wave model* of light, predict which should have more energy, bright red light or dim blue light. How should the energy of bright red light and bright blue light compare?

   b. Given a *particle model* of light, predict which should have more energy, bright red light or dim blue light. How should the energy of bright red light and bright blue light compare?

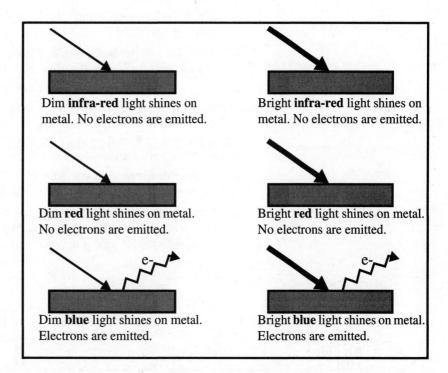

Figure 1-9:   Selected results of the photoelectric effect.

2. Examine the data in Figure 1-9 above and write down at least two observations.

3. From the data in Figure 1-9, which is more critical in getting electrons to leave the metal, the brightness (intensity) of light or the frequency of light? Explain.

4. Recall that removing electrons from a metal surface in the photoelectric effect requires light energy. What do the data in Figure 1-9 suggest about which is higher in energy, red light or blue light? Explain.

5. Do these results of the photoelectric effect provide support for the wave theory or particle theory of radiation? Briefly explain your reasoning.

Applying Your Ideas

6. Einstein used a particle model of light to explain the photoelectric effect. He proposed that light consists of particles of light energy called **photons.** Each photon carries a specific energy (in Joules, J) given by the relationship $E = h\nu$, where h is called Planck's constant and has the value 6.626×10^{-34} J · s, and ν is frequency of radiation in units of s^{-1} or Hertz. If a photon with energy *above* a certain minimum or threshold value strikes a material, electrons will be ejected. Do you see evidence for such a threshold energy value in the data in Figure 1-9? Explain your reasoning.

7. We can use the relationship, $E = h\nu$, to determine the energy per photon of radiation with frequency ν.

   a. Use this relationship to find the frequency of blue light with energy per photon 4.42×10^{-19} J.

   b. Recall the relationship between frequency and wavelength from Exploration 1B, $\nu = c/\lambda$. What is the wavelength of this blue light?

   c. What is the energy of a *red* photon with a wavelength of 650 nm?

8. The threshold energy required to remove an electron from a solid material is called the **work function**, Φ. Work function values are given in Figure 1-10 for elements that are solids at room temperature. The units are **electron volts (eV)**, where 1 eV = 1.6022×10^{-19} Joules.

   a. What general trends do you observe in Φ as a function of the location of the element in the periodic table? (Look down a column or across a row.)

   b. If the energy of incoming radiation is *greater* than the work function, electrons will be emitted from a material. Which element(s) will emit electrons in response to blue light if the energy of a blue photon is 4.42×10^{-19} J?

Li 2.9	Be 4.98											B 4.45	C 5.0				
Na 2.75	Mg 3.66											Al 4.28	Si 4.85				
K 2.30	Ca 2.87	Sc 3.5	Ti 4.33	V 4.3	Cr 4.5	Mn 4.1	Fe 4.7	Co 5.0	Ni 5.15	Cu 4.6	Zn 4.9	Ga 4.2	Ge 5.0	As 3.75	Se 5.9		
Rb 2.16	Sr 2.59	Y 3.1	Zr 4.05	Nb 4.3	Mo 4.6		Ru 4.71	Rh 4.98	Pd 5.1	Ag 4.6	Cd 4.2	In 4.12	Sn 4.4	Sb 4.6	Te 4.95		
Cs 2.14	Ba 2.7	La 3.5	Hf 3.9	Ta 4.25	W 4.55	Re 5.75	Os 4.8	Ir 5.7	Pt 5.65	Au 5.35	Hg 4.49	Tl 3.8	Pb 4.25	Bi 4.2			

Figure 1-10: The work function (Φ) for selected elements that are solids at room temperature. This parameter is a measure of how difficult it is to remove an electron from an element via the photoelectric effect, with higher numbers representing elements from which it is more difficult to remove an electron. (Source: *CRC Handbook of Chemistry and Physics*, 70th edition (CRC Press, Boca Raton, FL 1989), pp. E93-E94.)

9. The CCD (charge coupled device) detector, used for detecting starlight, is made of semiconductor materials such as silicon (Si) and germanium (Ge). It works by a process similar to the photoelectric effect. When light impinges upon the CCD, electrons move and will collect in certain regions of the device, *if the incident radiation is of a minimum energy*. In semiconductors, this minimum energy is called the **band gap**. Like the work function in metals, the band gap differs between semiconductors. Table 1-2 presents several semiconductors with their respective band gaps to consider using in a CCD.

    a. A semiconductor material can detect light with energy *greater than or equal to* its band gap. Could any of the materials in Table 1-2 below be used to detect red light of 650 nm?

    b. Will any of these materials be able to detect IR radiation with wavelength of 2000 nm (2 μm)? Explain your reasoning.

Table 1-2   Semiconductor Materials

Crystal	Band Gap (eV)
Si	1.11
Ge	0.66
GaP	2.25

Making the Link

Looking Back: What have you learned?

Chemical Principles

During Session 1 you learned important chemical principles and terms associated with electromagnetic radiation. You should be familiar with the following principles:

- Electromagnetic radiation: wavelength, frequency, amplitude, intensity, period. (Exploration 1B)
- Regions of the electromagnetic spectrum (Exploration 1B)
- Diffraction and interference (Exploration 1C)
- The photoelectric effect, photons, E = hν (Exploration 1D)
- Electrons, protons, atomic number, mass number, isotopes (Exploration 1D)

Thinking Skills

As you worked through Session 1, you developed some general problem-solving and scientific thinking skills that are not specific to chemistry. These skills are valued by employers in a wide range of professions and in academia. You have gained experience with the following skills:

- Data observation and interpretation (Exploration 1A and 1C)
- Data analysis skills (Exploration 1D)
- Evaluating a model (Exploration 1C and 1D)

Checking Your Progress

The following problems will help you integrate the concepts you have learned in Session 1, and begin to make progress toward the culminating activity.

1. In this Session, you have learned about several properties of light. Now you will try to describe the nature of light.

    a. Summarize briefly your understanding of the wave and particle properties of light and which experiments reveal these properties.

    b. How would you describe the nature of light? Explain your reasoning.

2. Write a brief response (no more than one page) to the question, *What is starlight?*, using the concepts you have learned in this Session.

Session 2   What do star colors tell us?

Blackbody Radiation, Color, and Temperature

Exploration 2A

Are all red objects "red hot"?

Creating the Context

Astronomers are interested in knowing the temperature of a star because it provides clues about the star's composition, and the state and structure of matter present in the star. Since our goal is to determine what elements are present in stars, it will be useful to know more about how we measure stellar temperature. In Exploration 1A you looked at images of several astronomical objects, where you may have noticed many different colors. What is the source of the variation in color of stars? Is there any relationship between the color of a star and its temperature? The goal of this session is to address the question, *What do star colors tell us?*

In the early part of this century, the director of the Harvard College Observatory, E. C Pickering and his collaborators, Williamina P. S. Flemming, Antonia C. Maury, and Annie J. Cannon, began a large study of nearly a quarter of a million stars. The result of their survey was a stellar classification system. Under this scheme, stars are classified into seven lettered categories that describe the surface temperature of a star: O, B, A, F, G, K, M. Class O stars are hottest, while class M stars are coolest. Our Sun is a class G star, with a surface temperature of about 5,800 Kelvin. Table 2-1 below summarizes the stellar classes and a temperature range that defines each. The temperature is given in degrees Kelvin (K), which is related to degrees Celsius (C) by $T(K) = T(C) + 273.15$.

Table 2-1   Star Classifications According to Temperature

Surface Temperature (K)	Class
30,000	O
11,000-30,000	B
7,500-11,000	A
6,000-7,500	F
5,000-6,000	G
3,500-5,000	K
3,500	M

As a first step in understanding what the color of stars tells us, we will focus upon objects that **emit** or give off radiation, such as stars and heated solids, and also those that do not emit visible light. We will consider the color and temperature of these objects and answer the question, *Are all red objects "red hot"?*

Developing Ideas

In this activity we consider what happens to the color of various objects when we add a source of energy such as heat from a flame or electricity. We will watch several movies of heated solids that can be found on the Module Web Tools CD-ROM in Exploration 2A of Stars. Alternatively, your instructor will provide an internet address for this page to you.

For each heated solid:

1. Predict what will happen to the color of light *emitted* by the object as it is heated. (Use the blank grid below for your answers.)
2. What source of energy is used to heat the solid?
3. What emitted colors do you observe, and how do they change with temperature?

Object	Prediction	Energy Source	Color, Temperature Observation
Piece of Iron			
Lime (calcium oxide)			
nichrome wire			
tungsten wire			

4. Given your observations above, propose a general relationship between temperature and the colors and wavelengths of light an object emits as it is heated.

Why are red objects red?

Your instructor will show you images of the following red objects:

- A red apple at room temperature
- Mars ($T_{surface} = 310$ K)
- Betelgeuse ($T_{surface} = 3,500$ K)

5. We often refer to Mars as the red planet, and Betelgeuse as a red star. Betelgeuse emits visible radiation, while Mars does not. Compare the temperature and color of these objects. Are all red objects "red hot"? Explain your reasoning.

Exploration 2B

Why is Mars red?

Creating the Context

When you observe the colors of a star such as Betelgeuse and the planet Mars, you find that they both appear red. As you learned in Exploration 2A, however, Betelgeuse and Mars have vastly different surface temperatures. You also learned that while Betelgeuse emits visible radiation, Mars does not. Thus, there may be several reasons that both Mars and Betelgeuse are red. Since our goal in this session is to learn what the colors of stars may tell us about their temperature and composition, we will need to understand color in more detail. We begin by learning about several processes that can determine the color of objects and answer the question, *Why is Mars red?*

Light emitted from the Sun appears white because it contains all the colors of the visible spectrum. We see individual colors when this visible light interacts with various objects. One way light can interact with an object is **reflection**. When sunlight encounters an **opaque** (non-transparent) object, part of the light will move away from the object and be reflected. The reflected wavelengths of light determine the color you see. For example, if you shine white light on a green shirt, your eyes detect the green light that is reflected from it.

What happens to the rest of the visible light? In the case of an opaque object such as the green shirt, the remaining wavelengths are **absorbed** by the object. When light interacts with a *transparent* object, some of the light is absorbed and some passes through the object or is **transmitted**. A piece of transparent green glass for example, transmits green light but absorbs the other wavelengths. Thus, we conclude that the color we see are those that are *not* absorbed by an object but make it to our eyes.

The key to understanding why objects appear colored is to learn more about the processes of absorption, emission, reflection, and transmission of light. In this Exploration, you will examine these processes in more detail by working with filters and objects of different colors. These experiments will help you better understand light and color.

Developing Ideas

PART I: PRISMS, DIFFRACTION GRATINGS, AND RAINBOWS

In this experiment, you will use a diffraction grating or prism to spread white light into components of various wavelengths and create a rainbow shining on the wall or screen. Then you will use food coloring to make up solutions of various colors, and place these solutions between the white light and the rainbow on the wall.

Before you begin, let's explore why a rainbow is produced when light passes through a **prism**. (see sidebar) The speed of light is the same whether it is visible sunlight, X-ray radiation, or radio waves, *as long as the light is in a vacuum.* When light passes through a medium other than a vacuum, its speed decreases because it interacts with particles in the medium. The **refractive index**, n, of a substance is a measure of the speed of light in the substance (at speed v) relative to the speed of light in a vacuum (usually denoted c); or $n = c/v$. A pencil partially submerged in water appears bent at the interface of the air and the water because water and air have different refractive indexes, and light moves at a different speed through each.

The refractive index is also dependent on the wavelength of light. That is why when light shines through a prism, the shorter wavelength light (blue, for example)

slit
prism
red
light source
blue

is bent more than the longer wavelength light (red, for example). Since light of different wavelengths is bent differently as it passes through a prism, the white light is dispersed into a rainbow. A rainbow is also produced when light passes through a **diffraction grating**. You will have a chance to explore the rainbow light produced by a diffraction grating.

Work in small groups for this activity. Each group should have a diffraction grating, access to a bright white light source (like a slide projector), food coloring (or colored filters), and a glass container.

1. Place a diffraction grating in front of the light from a slide projector and observe the resulting rainbow on a wall or screen. This will be easier if the room is darkened.

2. Fill a rectangular glass container with water. Add a few drops of red food coloring and stir the solution. (Alternatively, you can use colored filters.)

3. What do you think will happen to the rainbow when you put the red solution in its path? Record your prediction in the table below.

4. Place the red solution between the slide projector and the diffraction grating and record what happens to the rainbow on the wall.

5. Repeat steps 1-4 for other colors and record your results in the table below.

Solution Color	Predictions	Observations
red		
orange		
green		
blue		

6. What is the relationship between solution color and the color(s) of light absorbed by the solution? What is the relationship between solution color and color(s) of light that is transmitted by the solution? Explain your observations.

PART II: COLORED FILTERS AND COLORED OBJECTS

In this activity, you will work in small groups to explore how objects appear when viewed through colored filters. Each group should have a set of colored filters (blue, green, orange, red) and colored objects.

7. For each filter, what color light do you expect to be absorbed or transmitted? Make these predictions in the table below. (Recall your results from Part I using colored solutions. The solutions are transparent to light, as are the filters.)

Filter Color	Absorbs	Transmits
Red		
Orange		
Green		
Blue		

8. Record the color of various objects. Look at all objects and *note the color of the object when viewed through the filter* in the table below. Note in particular which objects turn dark or almost black when viewed through the filter.

Object Color	Green Filter	Red Filter	Blue Filter	Orange Filter

9. Discuss your results with your group. What conclusions can you draw from your data about the relationship between the color of an object and the colors it absorbs, transmits, or reflects? Focus especially on the objects that appear black or dark through the filters. Refer to the relationships in the color wheel. (see sidebar)

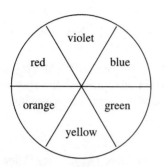

A color wheel shows the relationship of complementary colors. Each color is opposite its complement.

PART III: COLOR MIXING

In this exercise you will use a **Color from Emission** multimedia tool that will allow you to investigate the mixing of different colors of light. This tool can be found on the Module Web Tools CD-ROM in Exploration 2B and on the Stars web site. Alternatively, your instructor will provide the internet address for this page to you.

10. First start with an absence of light (black) by turning all the controls for emission of red, blue, and green to zero.

11. Predict what you think will happen when you add the amounts of light according to the table below (0 = no color, 1 = 100% of the color). Then observe and record the color produced.

Red	Green	Blue	Predicted Color	Observed Color
0	0	1	blue	blue
0	1	0	green	green
1	0	0	red	red
0	1	1	X yellow	teal
1	0	1	purple	purple
1	1	0	yellow	yellow
1	1	1		white

12. Yellow and orange are both colors found in the rainbow. How can you reproduce these colors by mixing red, green, and blue light?

    red and yellow

13. How do you get white by mixing red, green, and blue? How do you get various shades of grey?

    white includes all the colors of light. Red green and blue make all the colors of light. Lessen the intensity off all the colors, just keep them all the same

Applying Your Ideas

14. Make a sketch illustrating how light, objects, and your eyes interact in the following situations. Use appropriate words to briefly describe the processes (absorption, reflection, transmission) that produce the object color you observe. Be specific about the colors of light involved. Assume that the light source is white light from the sun.

   a.   A green, opaque object is seen by an observer.

   b.   A green, transparent object (such as a filter) is seen by an observer.

   c.   An green opaque object is seen through a red filter.

15. What color does a blue shirt, or other opaque object, appear when illuminated by blue light?   blue?

16. You saw in Part II that if you mix 100% red, 100% green, and 100% blue light, the result is white light. What color will result from mixing equal amounts of 100% red, 100% green, and 100% blue opaque paint? Explain your reasoning.

17. Why is Mars red? Draw a picture that shows: *yellow, the same amount of paint for all is used?*

   •   The source and type of light shining on Mars

   •   What happens to the light when it encounters Mars (what color(s) are absorbed, what color(s) are reflected)

   •   The light rays that reach the eyes of an observer (note the colors).

18. How would Mars appear if you could view it through a green filter? Explain your reasoning, noting the colors of light and processes involved (absorption, reflection, transmission).   *yellow, red and green together show yellow, red is reflected absorbed by filter transmitted as yellow*

Exploration 2C

Why is a star red?

Creating the Context

Two stars that have different colors are Betelgeuse and our Sun. Betelgeuse is a red star, while our Sun appears white. Why do we observe these star colors? Does the difference in star color suggest differences in other properties such as temperature and composition? Since our goal is to understand the composition of a star, in this Exploration we will learn more about the colors of stars and what they tell us and answer the question: *Why is a star red?*

In Exploration 2A, we saw light emitted from several heated solids, such as the tungsten filament in an incandescent light bulb. We saw that several colors of light are emitted, depending upon the temperature of the solid. The energy emitted as electromagnetic radiation by such objects is called **blackbody radiation**. Any *opaque* object such as a heated iron bar will emit blackbody radiation. Cool objects emit blackbody radiation as well, but primarily in a region of the electromagnetic spectrum that is invisible to our eyes. Although stars are not solids, we know they are composed of hot, dense gas. In fact, the stars we examined in Exploration 1A were not transparent, suggesting that stellar gas is highly opaque. Thus, as a *first model* of a star, we will propose a hot ball of gas that emits blackbody radiation. Using this model, we will be able to estimate the surface temperature of a star.

Preparing for Inquiry

A hypothetical perfect blackbody absorbs and emits all incident radiation, with its radiant emission dependent only on temperature. Although such a perfect radiator does not exist in nature, scientists use a blackbody model to approximate the way that objects such as heated solids emit light. By early 1900 scientists had struggled and failed to mathematically describe blackbody radiation. Then, in October 1900, Max Planck was able to theoretically model blackbody radiation, once he assumed that a blackbody radiates light in packets of energy called photons, with energy $h\nu$. Here, ν is the frequency of radiation and h is what we now call Plank's constant, 6.63×10^{-34} J·s. Recall from Exploration 1D that Einstein used this proposal of a particle nature of light in his explanation of the photoelectric effect.

Blackbody radiation occurs over a broad region of the electromagnetic spectrum, and the intensity can vary by many orders of magnitude (powers of ten) across the spectrum. Careful measurements of radiation spectra from hot objects show that as the temperature of the object increases, more light in the high-energy region (corresponding to shorter wavelength) will be found in its **spectrum.** A spectrum is a record of the amount of radiation as a function of wavelength or frequency. Figure 2-1 below shows the spectra of two blackbodies at different temperatures. On the y-axis is the **energy density** (energy per volume) per wavelength of the radiation emitted by a blackbody in units of Joules/cm^4.

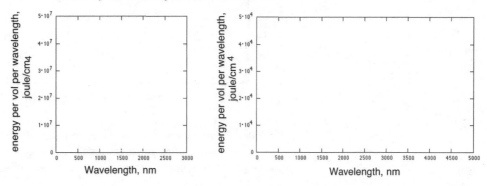

Figure 2-1:   Blackbody radiation curves for two different temperatures are shown, with the energy radiated per unit volume on the y axis and the wavelength on the x axis. Can you tell which curve represents the white-hot object and which represents the cooler red-hot object? Note that the y-axis scales are different in the two graphs.

The **intensity** of the radiation emitted by a blackbody is equal to $\sigma \cdot T^4$, where σ is the Stefan-Boltzmann constant, $\sigma = 5.67 \times 10^{-8}$ W·m^{-2}·K^{-4}, and T is temperature on the Kelvin scale. A watt is the amount of energy produced per unit time ($1\text{W} = 1$ J · s^{-1}) and the *intensity* in W/m^2 is the energy per second per unit area. For example, when the Sun shines on the earth, the incident solar energy on the earth's surface is typically calculated in Watts per square meter.

Developing Ideas

Blackbody Radiation and Temperature

In the following activity, you will explore how blackbody radiation curves change with temperature using the **Blackbody Radiation** tool on the Module Web Tools CD-ROM in Exploration 2C or on the web site. Alternatively, your instructor will provide the internet address for this page to you.

1.   Recall star classification scheme shown in Table 2-1. Betelgeuse is a class M star, while the Sun is a class G star. Which is hotter, Betelgeuse or the Sun?

2. What is the structure of a star? As a *first model* of a star, we will propose that it is like a blackbody. Thus, like a blackbody it should emit radiation over a broad range of wavelengths. To see what the spectrum would look like under this model for stars such as Betelgeuse and the Sun, click on the *Blackbody emission from 3,000 to 7,000 K* spectrum. Use the slide bar to locate the spectra corresponding to blackbodies at 5,800 K and 3,500 K. Roughly sketch both spectra in the space provided below.

    •   T = 5,800 K (the Sun)

    •   T = 3,500 K (Betelgeuse)

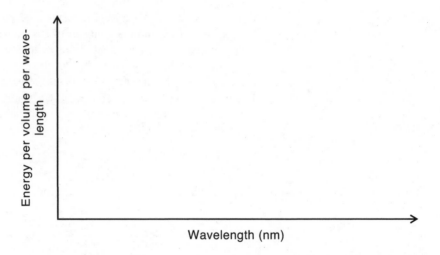

3. Write down two generalizations about how the radiation curves change as you increase the temperature. In particular, note changes in the shape, overall intensity, and the wavelength corresponding to maximum intensity (λ_{max}).

4. Approximate the λ_{max} of each blackbody spectrum. What region of the electromagnetic spectrum does the λ_{max} of each blackbody lie in? What color, if any, does λ_{max} correspond to?

Temperature (K)	λ_{max} (nm)	Region of EM spectrum and color
5,800		
3,500		

5. Assuming that the blackbody spectrum at 5,800 K models that spectrum of the Sun, based upon the color of λ_{max} of this blackbody, what color should the sun appear?

The Color of Stars

For the following questions you will again use the **Blackbody Radiation tool** on the Module Web Tools CD-ROM in Exploration 2C or on the Stars web site. However, you will need to use the spectra that contain only the *visible* portion of the blackbody emission from 350 to 700 nm.

6. To understand why our Sun is white, we need to consider the mixing of different colors of light. Recall your results from the color mixing exercise in Exploration 2B. What happens when you mix together all colors of visible light?

7. Click on the *Visible portion of blackbody emission plot from 3000-6500K*, and heat the object to 3,500 K. Based upon the visible colors present at this temperature and your knowledge of color mixing, what color will this blackbody appear?

8. Next, again using the plot from 3000-6500 K, heat the object to 5,800 K, where the full visible region is included in its blackbody spectrum. What color, if any, does the λ_{max} correspond to, and what color will the object appear?

9. Recall your prediction of the Sun's color based on the λ_{max} of its blackbody radiation curve. Why isn't the Sun green? Explain your reasoning.

10. How accurate is a blackbody model of a star? Your instructor will show you the simulated blackbody spectrum at 5,900 K and the actual spectrum of a class G star like our Sun. Compare the two spectra. Does a blackbody radiation model explain the spectrum of this star? Explain.

Applying Your Ideas

11. Use the **Plank Radiation Applet** on Module Web Tools CD-ROM in Exploration 2C or on the Stars web site to answer the following questions for each star in the table below. Alternatively, your instructor will provide the internet address for this page to you.

    a. Determine the λ_{max} in Angstroms (1 Angstrom = 10^{-10} meters) associated with a blackbody of the given temperature. Record this in the table below.

    • To do this: Select a temperature by dragging the side color bar to the desired temperature (or as close to it as you can get). Measure the wavelength (x-axis) by using the mouse to direct the cursor to a data point, and then click and hold down the mouse. A text box will appear with the wavelength in Angstroms. The intensity (y-axis) is given in intensity units.

    b. In what region of the electromagnetic spectrum does λ_{max} lie? What color, if any, would λ_{max} be? Record this in the table below.

Average T (K)	Class	Star	λ_{max} (Angstroms)	Region of EM spectrum with λ_{max}, color of λ_{max}
8,000	A	Altair		
6,500	F	Procyon A		
5,300	G	Capella		
4,500	K	Arcturus		

12. What is the mathematical relationship between T and λ_{max}? Are these directly proportional or inversely proportional to each other? Explain.

13. This relationship between the temperature of a blackbody and λ_{max} is known as **Wien's Law**, where the proportionality constant is 2.898 mm · K. Use Wien's Law, $T = 2.898$ mm· K/λ_{max}, to determine λ_{max} (in Angstroms) for the following stars.

a.   Class M Star, Betelgeuse, T = 3,500 K.

b.   Class B Star, Regulus, T = 12,000 K.

c.   Class O Star, Zeta Orionis, T = 30,000 K.

14. Predict the color of a class O star such as Zeta Orionis by using the *Blackbody Radiation* tool found on the Module Web Tools CD-ROM in Exploration 2C of the Stars web site. Click on the *Visible portion of blackbody emission plot from 10,000-28,00K*. Play the movie so that the object is heated to 28,000K. What region does the λ_{max} lie in? What color might this object appear?

15. Rank these stars in order from coolest to hottest. The stars appear: orange, red, white, yellow, blue. Which color stars emit the highest energy (shortest wavelength) radiation?

Making the Link

Looking Back: What have you learned?

Chemical Principles

During Session 2 you learned important chemical principles and terms associated with color, temperature, and blackbody radiation. You should be familiar with the following principles:

•   Emission of light (2A, 2B, and 2C)

•   Absorption, reflection, and transmission of light (2B)

•   Opaque and transparent objects (2B)

•   Blackbody radiation, λ_{max} (lambda max), and Wien's law (2C)

Thinking Skills

As you worked through Session 2, you developed some general problem-solving and scientific thinking skills that are not specific to chemistry. You have gained experience in the following skills:

•   Data observation and interpretation (2A)

•   Experimental design (2B)

•   Data analysis skills (2B and 2C)

•   Proposing and evaluating a model (2C)

Checking Your Progress

The following problems will help you integrate the chemistry concepts you have learned in Session 2, connect your understanding to the story line, and begin to make progress toward the culminating activity.

1.   *Estimate* the λ_{max} of the G star in Figure 2-2 and determine the temperature using Wien's law from Exploration 2C. Then estimate the color of the star based

on its temperature. (You may need to use the **Blackbody Radiation** movies on the Module Web Tools CD-ROM from Exploration 2C to help estimate color.)

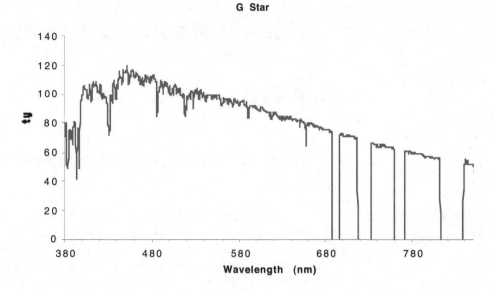

Figure 2-2: Spectrum of a G star taken from the holdings of the Astronomical Data Center (ADC), Silva, D., "A New Library of Stellar Spectra," 1992.

2. Compare the G star spectrum in Figure 2-2 to the A star spectrum in Figure 2-3.

  a. Which star is hotter? How do you know?

  b. Compare the appearance of these stellar spectra. What can you say about the spectra of stars at different temperatures? Explain.

  c. Can you explain this difference, based upon what you have learned so far? What else might you need to know?

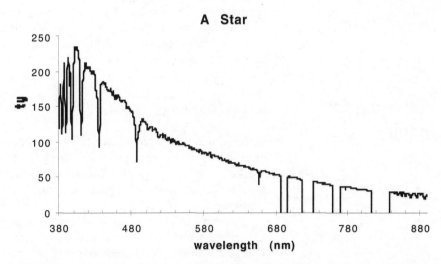

Figure 2-3: Spectrum of an A star taken from the holdings of the Astronomical Data Center (ADC), Silva, D., "A New Library of Stellar Spectra," 1992.

What do stellar spectra tell us? Part I

One-electron Atoms and Ions

Exploration 3A

How can we describe the spectrum of a star?

Creating the Context

In Session 2 you learned about the information revealed by the color of stars. You now know that the color of a star is determined by its temperature, and that the spectrum of a star can tell us about its temperature. However, we would still like to know what elements are present in stars. A logical next step is to see whether we can extract more information from the spectrum of a star that will provide clues to its composition. Thus, in this Session, we will re-examine stellar spectra and begin to answer the question, *What do stellar spectra tell us?*

In Exploration 2C, you learned that the spectrum of a blackbody is a broad, smooth curve, with maximum intensity at a certain wavelength determined by temperature. The smooth spectrum of a blackbody shows that many wavelengths of light are emitted, though perhaps not all with the same intensity. A star, such as our Sun, is an opaque object that emits radiation over a broad portion of the electromagnetic spectrum. Using this criteria, we proposed a blackbody model for a star.

However, in comparing the spectra of a blackbody and a star like our Sun, you saw that the actual spectrum of a star is not completely smooth. While the shape of the spectrum emitted by a star approaches a blackbody curve, there are wavelengths in the spectrum where light intensity is missing. These **fine features** of a stellar spectrum suggest that something is removing or *absorbing* light from the star at those wavelengths. In this Exploration we will think about what might be responsible for the fine features in a stellar spectrum and determine, *How can we describe the spectrum of a star?*

Preparing for Inquiry

A Model for Energy

So far we have considered the way that heated solids emit energy in our study of blackbody radiation. Our model for energy is one in which an object emits light as "packets" with energy $E = h\nu$. The object can emit many frequencies of light over a broad region of the electromagnetic spectrum, so that its radiation spectrum appears **continuous**.

Another model for energy that we may consider is a **discrete** model, in which energy still comes in packets, but only certain energies can be absorbed or emitted. Consider an analogy to two different ways of entering a building, either walking up stairs or walking up a ramp. With a ramp you must take a certain number of steps, but you can take steps of any size and still make progress toward the entrance. With stairs, however, you must take a certain number of steps, *and* you can only ascend in whole steps. (How much success will you have moving half a step at a time?) Using the stairs is an example of a *discrete* process, because only certain step sizes are possible.

Scientists often represent the wavelengths of light emitted or absorbed by an object through graphs. One common graphing technique is illustrated in Figure 3-1 (top). We will refer to this type of plot as an **intensity-wavelength spectrum,** because it shows how the intensity of light absorbed or emitted varies with wavelength. The blackbody curve we considered in Session 2 is an example of an intensity-wavelength spectrum. The dips in the horizontal line represent places where discrete wavelengths of energy are absorbed by atoms and molecules. These dips are called **spectral lines**.

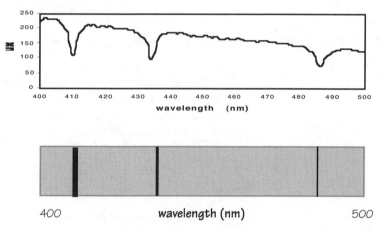

Figure 3-1:   Examples of an intensity-wavelength spectrum (top) and a line spectrum (bottom).

A second method shown in Figure 3-1 (bottom) is simply to plot vertical lines representing wavelengths where light is absorbed or emitted in a **line spectrum**. We'll use examples of both graphing techniques in this exploration.

A Model for Stellar Structure

In Exploration 2C, we described a star as a hot, opaque ball of gas that emits light as a blackbody. However, we soon discovered that this simple first model could not explain the spectral line features present in stellar spectra. Studies of our Sun, the star closest to us, yield a wealth of information about stellar structure and composition. Based upon such information from solar studies, we can now propose a *refined model* for the structure of a star's interior (Figure 3-2).

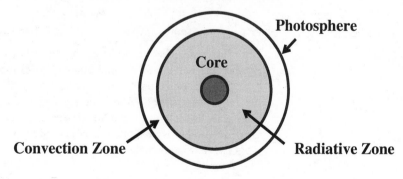

Figure 3-2:   A model for the structure of a star.

At the center of a star is the **core**, where energy is generated from the fusion of H nuclei. The core is extremely hot and dense; in the Sun's core, temperatures reach 16 million Kelvin! The **radiative zone** and **convective zone** are regions where energy is transported outward from the hot core to progressively cooler outer layers by photons and the motions of gas. The outer layer is the **photosphere**. It is analogous to the atmosphere of a star and is much less dense than the inner layers. The photosphere of a star is also many times cooler than the super-hot core. For example, in comparison to the Sun's core, its photosphere is a cool 5,800 K. As photons make their way through the photosphere and out into space, they encounter the gaseous atoms that compose this cool, diffuse layer. In this Exploration we will consider how the interaction of photons with atoms in the atmosphere of a star affect the spectrum we observe for that star.

Developing Ideas

Fine Features of a Stellar Spectrum

1. Figure 3-3 shows the spectrum of our Sun below and above earth's atmosphere compared with a blackbody radiation plot at 5900 K. In this activity we will focus on the **fine features** of the solar spectrum, where light intensity is "missing" at certain wavelengths as compared to the blackbody spectrum. What does it mean for light intensity to be "missing"?

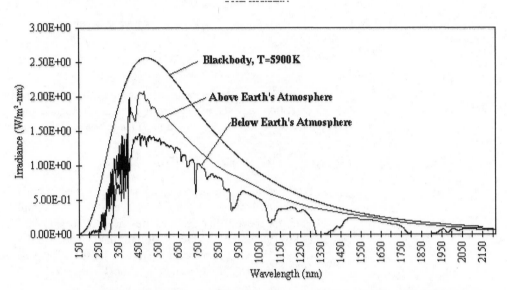

Figure 3-3:   The spectrum of the Sun above and below earth's atmosphere compared with a blackbody radiation curve at 5900 K.

2. Terrestrial gases, those in earth's atmosphere such as water and carbon dioxide, absorb radiation from the sun as it passes through the atmosphere. This absorption causes the prominent spectral lines with wavelength greater than 680 nm in the *Below Earth's Atmosphere* spectrum in Figure 3-3. However, these prominent features disappear in the spectrum taken above the earth's atmosphere. How might we explain other spectral lines that are not due to terrestrial gases? (Review the model of a star in Figure 3-2.)

How are atoms affected by light?

According to our model of a star in Figure 3-2, light energy originates in the core and passes through the photosphere, where gases can absorb some of the radiation before it reaches an observer on earth. Thus, in order to understand the fine features in stellar spectra and what they are telling us, we need to think about how substances absorb radiation. We will begin by observing the spectrum of hydrogen.

3. Observe the light from a hydrogen discharge tube. What color is it?

4. Observe the hydrogen emission with a diffraction grating slide and sketch the spectrum you observe in the space below.

```
┌──────────────────────────────────────────┐
│                                            │
│                                            │
│                                            │
│                                            │
└──────────────────────────────────────────┘
```

5. Now, observe an incandescent bulb (white light) with your diffraction grating slide. Make a sketch of the spectrum you observe in the space provided below.

```
┌──────────────────────────────────────────┐
│                                            │
│                                            │
│                                            │
└──────────────────────────────────────────┘
```

6. Compare your sketches from the light bulb and hydrogen discharge lamp. How would you describe each spectrum, as continuous or discrete?

7. We know that the spectrum of an incandescent light bulb can be explained by blackbody emission. But how can we explain the spectrum of hydrogen? We will focus on this task later in this Session. Consider what you already know from the photoelectric effect about light shining on a metal. What happens when light with energy above a certain threshold shines on atoms in a metal?

8. Now, think about shining light on a sample of gaseous atoms. What does the photoelectric effect suggest about one possible effect of shining light upon gaseous atoms?

Exploration 3B

What properties of atoms can help us learn about their structure?

Creating the Context

In Exploration 1D, you learned that shining radiation on metals will cause electrons to be emitted if the radiation is above a certain threshold frequency. This observation can be explained if we view light as composed of packets of energy called photons. The energy of each photon is proportional to the frequency, $E = h\nu$, where h is Planck's constant. You also observed that the minimum energy required to eject electrons from a solid varies, depending on the identity of the substance.

For gaseous atoms, there is a similar energy threshold for removing an electron, called the **ionization energy.** The value of this ionization energy (IE) depends upon the arrangement of electrons within the atom and differs between elements. Other properties of atoms, such as size and the energy associated with adding an electron, also depend on electronic structure. This suggests that we might learn something about the structure of different atoms by studying trends in atomic properties. Thus, the goal of this Exploration is to answer the question, *What properties of atoms can help us learn about their structure?*

According to our current model of a star, energy that originates as photons in the core of a star passes through an outer layer called the photosphere. This light eventually makes its way across space, through the earth's atmosphere, and to the observer on earth. However, you know from examining the spectrum of a star that not all photons of light pass through the photosphere; some are absorbed by elements in the photosphere. To help us begin to learn about how light interacts with atoms, in Exploration 3A we observed the emission spectrum of hydrogen. We saw that the emission spectrum consists of discrete spectral lines at specific wavelengths. Although the results of the photoelectric effect suggest that light affects the electrons in atoms, we cannot fully explain the discrete nature of the hydrogen spectrum. Thus, we need to learn more about how light affects the electrons in atoms, and how these electrons are arranged in the atom.

We will begin by observing some properties of individual atoms in the gas phase, since the results are easiest to interpret under these conditions. Trends we can observe as a function of location in the periodic table will help us learn more about the structure of atoms. A knowledge of atomic structure is essential for an understanding of how light interacts with atoms, and for satisfying our ultimate goal of identifying the elements in stars.

Preparing for Inquiry

How Much Energy is Required to Remove Electrons from Atoms?

The energy required to remove an electron from a gaseous atom to create a positively charged gaseous **cation** is known as the **ionization energy** (**IE**). Because stars are so hot, many of the atoms in the stellar atmosphere have lost one or more electrons. By convention, the ionization energy has a positive sign, because energy is required to pull a negatively charged electron away from the positively charged protons in the nucleus.

$$M_{(g)} + IE \longrightarrow M^+_{(g)} + e-$$

As the atomic number increases, the number of protons in the nucleus increases. To maintain neutrality, the number of electrons also increases. Figure 3-4 shows a plot of the **first ionization energy** (IE_1), the energy associated with removal of the most weakly held electron in an atom, as a function of atomic number. It is also possible to remove more than one electron from an atom. The energy required to remove the next most weakly held electron is called the second ionization energy (IE_2). Because stars are so hot, many of the atoms in the stellar atmosphere are missing one or more electrons. Table 3-1 lists the energies required to remove successive electrons for several elements.

The magnitude of the ionization energy provides information about the strength of attraction between the electron and the rest of the atom. A large IE means that an electron is difficult to remove, while a small IE means that an electron can be easily removed. Thus, we might be able to use the trends in IE across different elements in the periodic table to develop a way to think about the arrangement of electrons in an atom.

Table 3-1   Multiple Ionization Energies

Ionization Energy	O, Z = 8 (MJ/mol)	Na, Z = 11 (MJ/mol)	Si, Z = 14 (MJ/mol)
IE_1	1.3140	0.4958	0.7865
IE_2	3.3882	4.5624	1.5771
IE_3	5.3004	6.912	3.2316
IE_4	7.4693	9.544	4.3555
IE_5	10.9895	13.353	16.091
IE_6	13.3264	16.610	19.785
IE_7	71.3345	20.115	23.786

What Determines the Strength of Attraction between Electrons and the Nucleus?

We can better understand the energy of interaction between charged species such as electrons and protons by learning about Coulomb energy. The **Coulomb energy** is negative between particles of opposite charge (attractive interaction) and positive between particles of like charge (repulsive interactions). The energy of the interaction is proportional to the magnitude of the charge and inversely proportional to the distance between charges, as indicated by **Coulomb's law**:

$$E_{attraction} = \frac{kq_1q_2}{r}$$

where q_1 is the charge on one particle, q_2 is the charge on the other particle, r is the distance between them, and k is a proportionality constant. The constant k equals 9.0 x 10^9 N m^2/C, when the charge is in Coulombs (C), the distance is in meters (m), and the force is in Newtons (N).

According to Coulomb's law, two factors determine the strength of attraction between an electron and the nucleus: the magnitude of charge and the distance between the charges. The magnitude of charge between the nucleus and an atom is determined by the number of protons in the nucleus. In an atom, an electron carries a negative charge, with e (e = 1.621773 x 10^{-19} C). The nucleus carries a positive charge, which is often written as Ze, in terms of atomic number (Z) and electron charge (e). Thus, for the attraction between an electron and the nucleus, Coulomb's law above can be rewritten as $E_{attraction} = -kZe^2/r$.

Developing Ideas

Ionization Energy and Atomic Radii Trends

In this activity you will examine how the ionization energy and atomic radii of an element vary as a function of the element's position in the periodic table. As the atomic number of an element increases, the number of electrons increases by the same amount. We will use this information to begin building a model of where the electrons are in atoms.

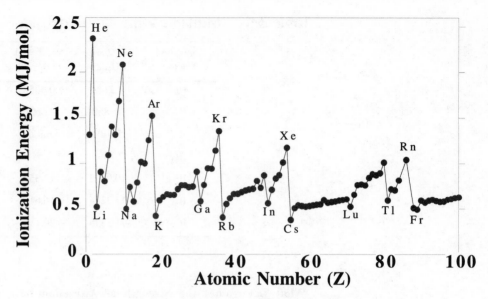

Figure 3-4:   A plot of the first ionization energy versus atomic number.

1. The elements in column 18 in the periodic table are known as the noble gases. Notice in Figure 3-4 that helium, one of the noble gas elements, has the highest first ionization energy. What does this mean in terms of how easy or difficult it is to remove an electron from helium compared to other elements?

2. What are the trends in the magnitude of first ionization energy as the atomic number increases *down a column* in the periodic table? *Across a row* in the periodic table?

3. What is the trend in atomic radii as atomic number increases *down a column* in the periodic table? (See Table 3-2.)

4. What is the trend in atomic radii as atomic number increases *across a row* in the periodic table? (see Table 3-2.)

Table 3-2   Radii of Selected Elements

Element	Atomic Radius (pm)	Element	Atomic Radius (pm[a])
Li	152	F	71
Be	113	Cl	99
B	88	Br	114
C	77	I	133
Na	186	O	66
Mg	160	S	104
Al	143	Se	117
Si	117	Te	143

[a]pm = picometers = 10^{-12} m

5. Summarize below the periodic trends you identified in IE and atomic radii:
    *   Across a row
    *   Down a column

6. According to Coulomb's law, one factor that determines the strength of attraction between two charged particles is the distance between them. The distance between the nucleus and an electron depends upon the size of the atom. A large atom's electrons will be further from the nucleus than a small atom's. Thus, it should require *less* energy to remove the electron furthest from the nucleus in a large atom than in a small atom. Do you find evidence for this in the periodic trends you identified in Problem 5? Explain.

Applying Your Ideas

7. Elements that require a great deal of energy to lose an electron are particularly *stable* in their atomic form, meaning they do not react with other substances readily. Using Figure 3-4, determine which column of elements in the periodic table has the *greatest* stability and which column has the *least* stability, and explain how you know this.

8. Examine the multiple ionization energies for O, N, and Si in Table 3-1. The first IE (IE_1) represents the formation of a +1 cation (O^+). IE_2 represents the energy needed to remove an electron from this +1 cation to make a +2 cation (O^{+2}). What patterns do you notice as you move from IE_1 to successively higher IE?

9. Energy changes also result when an electron is *added* to a neutral gaseous atom to form a negatively charged, gaseous **anion**. These energy changes are referred to as the **electron affinity (EA)** of an element. Although it always costs energy to remove an electron, adding an electron may either require or release energy. When energy is released, the EA is negative. When energy is required, however, the EA is positive. (Your instructor will inform you if your textbook has a definition of EA that differs slightly from this one.)

    a. Examine the plot of EA in Figure 3-5 below. Which elements have positive EAs? Which have negative EAs? What does a positive or a negative EA mean in terms of the ease or difficulty of adding an electron to these elements?

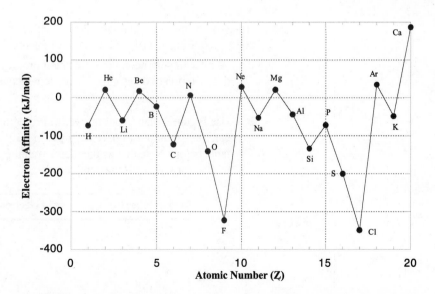

Figure 3-5:   Electron affinity values for the first 20 elements. Values are taken from J. Emsley, *The Elements*, third edition, Oxford University Press, 1988.

Exploration 3C

What does ionization energy tell us about the structure of atoms?

Creating the Context

Our goal is to understand how light interacts with atoms in the photosphere of a star, and how this interaction produces the fine features in a stellar spectrum. In Exploration 3B you began to think about where the electrons are in an atom by examining trends in several atomic properties. These trends verify Coulomb's law and tell us that the lowest energy and easiest to remove electrons are farther away from the nucleus. However, Coulomb's law does not tell us the *particular* arrangement of electrons around the nucleus. Thus, in this Exploration, we will examine the more subtle details of the ionization energy trends and answer the question, *What does ionization energy tell us about the structure of atoms?*

Although you may already have seen models for the structure of an atom, we want to go back and trace these ideas from their origin. We want to develop a model of the atom by using evidence found by interpreting experimental data. A closer look at ionization energy data will help us develop such a model and will provide a framework for understanding how light interacts with atoms. Ultimately, we will use our knowledge of atomic structure to extract information from the fine features of stellar spectra about the atoms in stars.

Developing Ideas

Shell Model

One model of the atom is the **shell model**, in which electrons reside in layers or shells around the nucleus, something like an onion, each ring of which represents a different shell. It is common to label these shells with integers **n**, beginning with n = 1 as the shell *closest* to the nucleus. Electrons first fill shell n = 1, then shell n = 2, etc.

1. It has been proposed that an element with *electron-filled* shells is particularly stable, requiring a very large amount of energy to remove an electron.

    a. Work with your neighbor or in small groups. Use the data in Figure 3-6 and this criterion to determine which elements have filled shells.

    b. Use Figure 3-6 and your answer to Problem 1a to determine the number of electrons required to fill each shell in an atom. Record your results in the table below. Note that as the atomic number increases, so does the number of electrons, by the same amount.

Shell number (n)	# of electrons in filled shell
1	
2	
3	
4	

Subshells

2. Again, examine Figure 3-6. What more subtle patterns do you notice in the trend in ionization energy as atomic number increases between Li and Ne? Between Na and Ar? Between K and Kr?

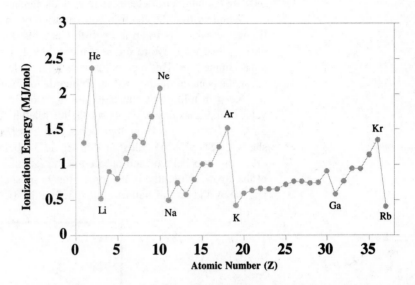

Figure 3-6: A plot of the first ionization energy versus atomic number of the first 40 elements.

3. These patterns suggest that we should include **subshells** within a shell in our model. As with filled shells, a *filled subshell* configuration presents a particularly stable arrangement of electrons. Thus, an element with a filled subshell will require more energy to remove an electron.

    a. Use Figure 3-6 to record in the table below the number of electrons that fill each subshell within the given shell

    b. Historically these subshells are labeled with ns, np, nd, and nf (n = shell number), according to the number of electrons that fill the subshell. The **s** subshell is filled by 2 electrons, the **p** subshell by 6 electrons, and the **d** subshell by 10 electrons. Use this notation to label the subshells in the table. (Note: The p subshell is filled by 6 or 3+3 electrons.).

Shell # (n)	Number of electrons in filled subshell	Subshell labels
1	2	
2		
3		
4		

Electron Configurations

Now we can use the shell model to describe where electrons are in an atom, using a notation known as **electron configuration**. Each element is given a label containing the numbers of the primary shells occupied (1, 2, 3, etc.) and the letters associated with the occupied subshells (s, p, d, etc.). For example, the electron configuration label of hydrogen is $1s^1$ and that of oxygen is $1s^2 2s^2 2p^4$, where the superscripts indicate the number of electrons in each designated subshell.

When writing the electron configuration of an element, we imagine that we are filling the subshells in an atom with electrons. First we fill the subshell of lowest energy, then we move to the next higher subshell, until all the element's electrons are accounted for. This method of working from lowest energy toward higher-energy subshells is based on the **Aufbau Principle** (aufbau is German for "building up").

Keep in mind that subshells are not exclusively filled in order of increasing value of n. For example, the 4s subshell is filled before the electrons are added to the 3d level. There are several approaches to remembering the energy order of the subshells. One approach is illustrated in Figure 3-7 below. Begin the order with the lowest energy subshell, 1s, and proceed to the 2s, 2p, 3s, etc. by following the direction of the arrows. Note that in this order, the 4s subshell is filled before the 3d subshell, and the 6s and 5p subshells are filled before the 4f.

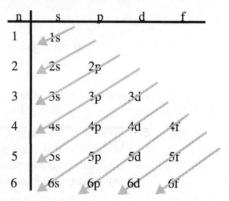

Figure 3-7:   A diagram that summarizes the order in which subshells are filled in an atom.

4.   Write the electron configuration for He, Ne and Ar. What do the electron configurations of elements in the same column of the periodic table seem to have in common?

5.   According to the shell model, the shell with the highest n number is *furthest* from the nucleus. Consider the electron configurations of He, Ne and Ar. Use your understanding of the shell model and Coulomb's law to rationalize the IE trends for these three elements.

6.   Sketch the shell structure of He, Ne, and Ar, showing labeled subshells, appropriate numbers of electrons in each subshell, the relative distance of each shell from the nucleus, and the charge of the nucleus (Z). He is done as an example for you below.

Applying Your Ideas

7. Each curved line in Figure 3-8 below represents a subshell within a shell of oxygen and O^+ ($Z = 8$). Each subshell is labeled using the ns, np, nd notation. According to the shell model, the $n = 1$ shell is closest to the nucleus, the $n = 2$ shell is the next closest, etc. Assume also that the s subshell is closer to the nucleus than the p subshell. (In Exploration 4E you will see evidence for these assumptions.)

   a. Write the electron configurations for O and O^+.

   b. Fill each subshell of O with electrons according to the Aufbau Principle and your electron configuration.

   c. According to Coulomb's law and the relative distance from the nucleus of each subshell as shown in the figure below, which subshell of O is highest in energy?

   d. Fill each subshell of O^+ with the appropriate number of electrons.

   e. We often refer to the shell (n number) farthest from the nucleus as the **valence shell**, and the electrons in this shell as **valence electrons**. How many electrons are in the valence shell of O and O^+?

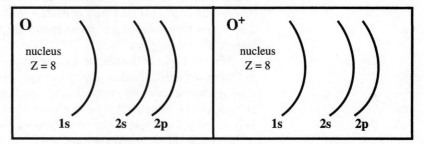

Figure 3-8: A diagram representing subshells in oxygen and its ion, O^+.

8. Examine Table 3-1 of multiple ionization energies for oxygen (see page 31). The electrons in the other shells closer to the nucleus are called **core** electrons. According to this table, what can you say about the energy required to remove valence electrons as compared to core electrons in oxygen?

9. Write the electron configuration for each element in the following sets.

   a. Li, Be, B

   b. F, Cl, Br

   c. Na, Mg, Al

10. It has been proposed that in addition to filled subshells, **half-filled** subshells represent a stable configuration. Use Figure 3-6 and your knowledge of the number of electrons in a subshell to find data that supports this hypothesis.

Exploration 3D

How does starlight rearrange the electron of hydrogen?

Creating the Context

In Exploration 3C, we examined the *removal* of electrons from atoms as a means to probe their structure. Our study of ionization energies provided a model for the atom in which electrons are arranged in shells and subshells around the nucleus, similar to the layers of an onion. We saw that the shell model and Coulomb's law explain why some electrons are held more tightly than others, with a general *increase* in the ionization energy of the electron as you proceed across a row in the periodic table and a general *decrease* in ionization energy of the electron as you proceed down a column.

Ionization is a fairly high energy process that causes complete removal of an electron from an atom. It is possible, however, to use energy to temporarily rearrange the electrons in an atom and move them between different shells. In Exploration 3A you saw an example of this process in the emission spectrum of hydrogen, where each spectral line represents the light emitted when the electron moves from a shell of higher energy to one of lower energy. The discrete nature of this spectrum implies that only certain wavelengths of light interact with hydrogen. However, many questions still remain, such as How are emission and absorption of energy related to the shell structure of atoms? Why are only certain energies emitted or absorbed? How is absorption different from emission in terms of electron rearrangement? And how and why do elements differ in their spectra? In this exploration we will develop a quantitative means of evaluating the energies absorbed and emitted by a hydrogen atom, and we use this to answer the question, *How does starlight rearrange the electron of hydrogen?*

We will begin by looking at the spectrum of the hydrogen atom. Not only is hydrogen the most abundant element in the universe and therefore likely to be in stars, it also possesses the simplest spectrum. You already know that the electrons in an atom can be thought of as being arranged in shells about the nucleus. Using this knowledge and a simple mathematical model, we will be able to predict how light interacts with the hydrogen atom and test our shell model of the atom. We will also use the discrete wavelength of light emitted and absorbed by hydrogen to identify its presence in stars.

Preparing for Inquiry

The Bohr Model of the Atom

One theory to explain the discrete wavelengths observed as emission lines was proposed by Niels Bohr in 1913. As spectroscopists were measuring light emitted by a variety of energetically excited chemicals, they noted that gaseous atoms and molecules that had been excited by light or electrical energy tended to emit light as sharp lines, rather than as the continuous blackbody radiation spectrum observed with opaque objects. For example, hydrogen, neon and mercury vapor have characteristic colors when they are electrically excited by a high voltage, and their spectra are a series of lines. The distinctive spectrum of each element serves as a kind of fingerprint; that is, each gaseous element produces a characteristic, unique line spectrum.

So far as we know, elements and their spectra are the same everywhere in the universe, enabling astronomers to detect these chemical species with their powerful telescopes and spectrometers. The element helium was, in fact, first detected through its unusual lines in the Sun's spectrum. Subsequently it was identified on earth, and the match in spectral lines was regarded as definitive evidence that the two observations were of the same species.

In his model of the atom, Bohr proposed that an electron in an atom could only possess certain discrete energies associated with shells or *energy levels*, numbered in integers from n =1 through n = infinity (∞). He called these allowed energy levels **stationary states** of the atom. According to Bohr's model, the energy of the electron comes in discrete amounts or is **quantized**. One example of a quantized object is chocolate. You could buy a whole bar, but not half a bar of chocolate. Likewise, the mass of chocolate is in fixed increment amounts. If the mass of a single uneaten bar is 5 grams, then you could buy 10 grams, 15 grams or 20 grams of chocolate, but never 2.5 grams.

Bohr used a planetary picture to visualize the stationary states in an atom. He viewed the stationary states as **orbits** of quantized energy, in which electrons move about the nucleus much the same way the planets move about the Sun. Although we will see that Bohr's picture of electron orbits is incorrect, his idea that the energy of an electron in an atom is quantized is both correct and essential to the modern theory of the atom.

Atomic Absorption and Emission

Figure 3-9 shows two of the stationary states in a sodium atom. These energy levels have energies of -5.15 eV (-8.25x10^{-19} J) and -3.04 eV (-4.87x10^{-19} J), *below* the energy required to ionize the atom ($E_{ionization}$), which is arbitrarily assigned the value of zero. The stationary state with the most stable, lowest-energy arrangement of electrons is called the **ground state** (E_{ground}). Each higher energy level is an **excited state** ($E_{excited}$), in which electrons are rearranged away from the lowest energy configuration. If a sodium atom absorbs energy and is excited to $E_{excited}$, the atom will *emit* energy and make a **transition** back to the more stable E_{ground} as indicated by the arrow in Figure 3-9. Often, the energy emitted by an atom is in the form of light. Sodium emits intense light at 598 nm as a result of electron transitions between two of its energy levels. This light lies in the visible (yellow) region of the electromagnetic spectrum and is responsible for the yellow color you associate with certain street lamps.

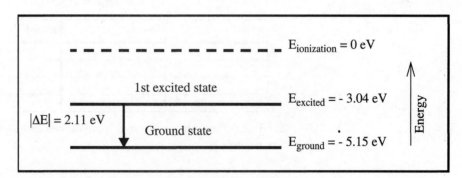

Figure 3-9: Select energy levels of the sodium atom.

Hydrogen is one of the most abundant elements in the universe. Hydrogen atoms can be prepared in the laboratory from hydrogen molecules, H_2, by means of an electrical discharge:

$$H_2 \text{ (g)} \longrightarrow 2 \text{ H (g)}.$$

Bohr's great insight was to show that each line in the spectrum of hydrogen corresponded to the electron's moving from one quantized energy level or stationary state to another. When a hydrogen atom *absorbs* energy in the form of light, heat, or electricity, the electron is excited to a *higher* energy level. Conversely, when the atom *emits* energy, usually in the form of light, the electron moves to a *lower* energy level (see Figure 3-10 below).

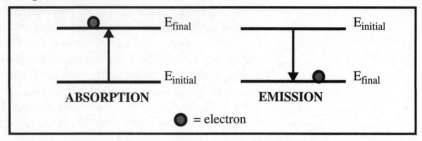

Figure 3-10: A representation of absorption and emission in hydrogen atom.

The absolute energy difference between the two levels, $|\Delta E| = |E_{final} - E_{initial}|$, corresponds to the photon wavelength we observe according to:

$$|\Delta E| = h\nu = hc/\lambda$$

As we saw in Exploration 3A, we can plot these discrete wavelengths in a line spectrum. We often refer to the spectral lines as **spectral transitions**. Using the observed spectral transition wavelengths for hydrogen, Bohr found that the energies of the levels in hydrogen atom were given by:

$$E_n = (-2.178 \times 10^{-18} \text{ J}) Z^2/n^2$$

In this equation, n represents the energy level numbered from n = 1 to infinity, Z is the atomic number, and energy is in Joules. In this activity you will construct an energy level diagram for hydrogen using the Bohr equation above and observed spectral lines of the hydrogen spectrum.

PRE-LAB QUESTIONS

1. Given the expression for E_n above, with Z=1, it is possible to calculate the energy of each energy level of hydrogen, starting with n=1. Calculate the energies of the 8 lowest energy levels, and enter these values in Table 3-3 below.

Table 3-3

n level	Energy (J)	n level	Energy (J)

2. Sketch the shells n=1 to n=8 for hydrogen and fill the appropriate shell with its electron.

3. Another way to visually represent the structure of an atom is through an **energy level diagram.** A simplified example of one is shown for sodium in Figure 3-9. On a separate sheet of paper, begin an energy level diagram by drawing each of the energy levels (n) in Table 3-3 as a short horizontal line. Start with the lowest n level at the bottom of the page, and draw higher n levels above it. The separation between each n level is determined by the energy difference between each level. Try to draw your diagram with these relative energy differences in mind. Beside each n level, write the corresponding energy (in Joules).

Developing Ideas

PART I: MEASURING THE SPECTRUM OF HYDROGEN

When you look at a hydrogen emission lamp through a spectroscope, the light is spread out into its individual wavelength components by a diffraction grating. You saw this effect in Exploration 2B when you used a diffraction grating to disperse white light into a rainbow.

4. Table 3-5 shows several spectral lines of hydrogen. Can you detect any of these wavelengths with your eyes? Explain.

5. Use the spectroscope to measure the visible wavelengths of the spectral lines of the hydrogen spectrum from 400 nm to 700 nm. (You should find 3-4 lines at approximately 656, 486, 434, 410 nm).

6. Are these lines that you observe due to absorption or emission of light by hydrogen atoms? What is happening to electrons at the atomic level to form these spectral lines? Draw a picture to illustrate this (see Figure 3-10).

PART II: ENERGY LEVELS OF HYDROGEN

7. The lines in the hydrogen spectrum all arise from transitions made by the electron from one energy level to another.

    a. Calculate $|\Delta E|$ in Joules, for the transitions that can arise from any two of the levels of hydrogen indicated in Table 3-4. Use the values for E_n you calculated in Table 3-3. Several answers are given below.

    b. Then determine the wavelength in nanometers using $|\Delta E| = hc/\lambda$, and enter these values in Table 3-4.

Table 3-4 Calculated Spectral Wavelengths

	$n_i = 8$	$n_i = 7$	$n_i = 6$	$n_i = 5$	$n_i = 4$	$n_i = 3$	$n_i = 2$
	$\|\Delta E\|$ (J), λ (nm)	$\|\Delta E\|$ (J), λ (nm)	$\|\Delta E\|$ (J), λ (nm)	$\|\Delta E\|$ (J), λ (nm)	$\|\Delta E\|$ (J), λ (nm)	$\|\Delta E\|$ (J), λ (nm)	$\|\Delta E\|$ (J), λ (nm)
$n_f = 1$	2.144×10^{-18} J 92.82 nm						
$n_f = 2$		5.001×10^{-19} J 398.0 nm					
$n_f = 3$			1.815×10^{-19} J 1096 nm				
$n_f = 4$				4.901×10^{-20} J 4061 nm			
$n_f = 5$							
$n_f = 6$							
$n_f = 7$							

PART III: ASSIGNING SPECTRAL LINES OF HYDROGEN

8. Record the wavelengths you measured in Part I in the blank spaces in Table 3-5 below. Compare the wavelengths you calculated in Table 3-4 above with the ones in Table 3-5. Your calculated wavelengths should match several of the observed wavelengths. Complete Table 3-5 by assigning each spectral wavelength to the upper and lower energy levels ($n_{initital}$ and n_{final}) involved in each transition.

Table 3-5   Hydrogen Spectral Line Assignments

Wavelength (nm)	Assignment $n_{initial} \dashrightarrow n_{final}$	Wavelength (nm)	Assignment $n_{initial} \dashrightarrow n_{final}$	Wavelength (nm)	Assignment $n_{initial} \dashrightarrow n_{final}$
		102.6		1004.9	
		121.6		1093.8	
		388.9		1281.8	
		397.0		1875.1	
97.3		954.6		4050.0	

9. Complete your energy level diagram for hydrogen by drawing a vertical arrow in the direction from $E_{initial}$ to E_{final} between pairs of energy levels associated with each transition in Table 3-5.

PART IV: THE BALMER SERIES

10. The **Balmer series** in the hydrogen spectrum consists of a series of lines in the visible region from 300 nm to 700 nm. From your entries in Table 3-5, what common characteristic do the Balmer lines in an *emission* spectrum have? In an *absorption* spectrum?

11. Most hydrogen atoms are in their ground state at typical temperatures, that is, the electron is in the n=1 level. For you to see Balmer absorption lines, the electron must originate in a higher energy level, the n=2 level. How much energy is required to move an electron from the n=1 ground level to the n=2 level?

Applying Your Ideas

The spectrum of a class A star in Figure 3-11 is an example of an *absorption* spectrum, plotted as intensity versus wavelength. An absorption spectrum is produced when starlight containing photons of all wavelengths is sent through gaseous atoms in the stellar atmosphere. We measure the transmitted light and find that photons at specific wavelengths are missing and have been absorbed by atoms in the sample. The corresponding dips in light intensity in the spectrum are the spectral lines.

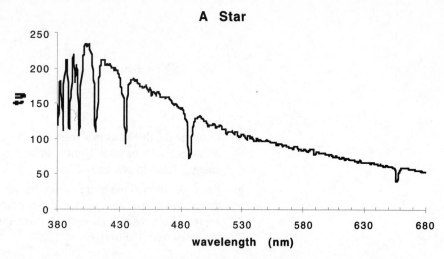

A Star

Figure 3-11: The spectrum of a class A star, plotted from 370-670 nm, from the holdings of the Astronomical Data Center (ADC), Silva, D., "A New Library of Stellar Spectra," (1992).

12. Measure and record the wavelengths of all spectral lines from 400 nm to 660 nm in the expanded portion of Figure 3-11 shown below. What element is responsible for these absorption features? Explain your reasoning.

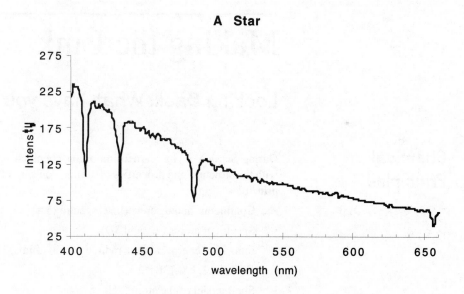

A Star

13. In 1896, E. C. Pickering found a series of unidentifiable lines in the spectrum of a class O star (T ~ 30,000K). The pattern of lines looked similar to that of the hydrogen Balmer series, but each line was shifted in wavelength by the same amount. Bohr identified these lines as belonging to $He^+(Z = 2)$ and showed that his equation could predict the wavelengths of these electron transitions.

a. Calculate the wavelengths of He^+ emission lines with $n_{final} = 2$, and $n_{initial} = 3$ and 4. What type of electromagnetic radiation is this? How do these wavelengths compare to transitions between corresponding levels in hydrogen?

14. Examine the transitions a, b, and c in the energy level diagram below.

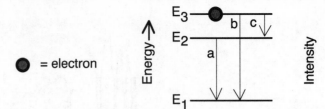

= electron

a. Examine the three electron transitions in the energy diagram above. Which transition, a, b or c, is highest in energy? Which transition is lowest in energy? How do you know?

b. Each transition in the energy level diagram corresponds to a peak or line in the emission spectrum below. Label each peak in the spectrum with the appropriate transition (a, b, or c). Note that energy in the spectrum below increases from left to right.

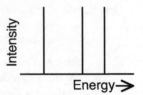

Making the Link

Looking Back: What have you learned?

Chemical Principles

During Session 3 you learned important chemical principles and terms associated with the structure and properties of atoms. You should be familiar with the following principles:

- Continuous and discrete (line) spectra (3A)
- Spectra of atoms (3A and 3D)
- Ionization energy, atomic radii, electron affinity (3B)
- Coulomb's law (3B)
- Shell model of the atom (3C)
- Electron configuration, Aufbau Principle (3C)
- Valence shell and valence electrons (3C)
- Bohr model of hydrogen atom and quantization of energy (3D)

Thinking Skills

As you worked through Session 3, you developed some general problem-solving and scientific thinking skills that are not specific to chemistry. You have gained experience with the following skills:

- Data observation and interpretation (3A, 3B and 3C)
- Data analysis and graphing skills (3B, 3C, and 3D)

- Proposing and evaluating a model (3A and 3D)
- Using experimental techniques (3D)
- Interpreting and using mathematical formulas (3D)

Checking Your Progress

The following problems will help you integrate the chemistry concepts you have learned in Session 3, connect your understanding to the story line, and make progress toward the culminating activity.

1. **Ions in Stellar Spectra.** Measure the wavelengths of the labeled peaks in the stellar spectra in Figure 3-12 and Figure 3-13. Use the list of selected ionic spectral lines in Table 3-6 below to identify the peaks.

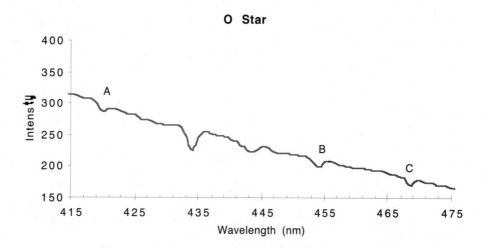

Figure 3-12: Spectrum taken from the holdings of the Astronomical Data Center (ADC), Silva, D., "A New Library of Stellar Spectra" (1992)

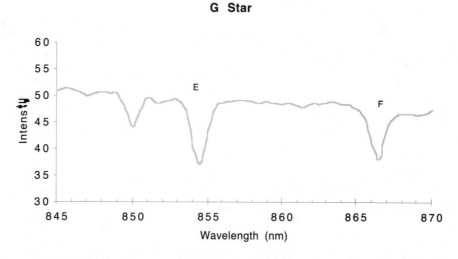

Figure 3-13: Spectrum taken from the holdings of the Astronomical Data Center (ADC), Silva, D., "A New Library of Stellar Spectra" (1992).

Table 3-6   Spectral Lines for Selected Ions

Ion	Wavelength (nm)	Ion	Wavelength (nm)
He^+	420.0	Ca^+	396.9
He^+	454.1	Ca^+	393.4
He^+	468.6	Ca^+	849.8
He^+	541.2	Ca^+	854.2
C^+	426.7	Ca^+	866.2
Mg^+	448.1	Si^+	412.2

2. **How are ions formed in stars?** You know that forming an ion requires energy, called the ionization energy. Some sources of energy that you may be familiar with are electrical energy, heat energy, and light energy. A source of energy available in stars is **thermal energy**, which is associated with the motions and collisions of atoms in a gas and is closely connected with temperature. The thermal energy of a star increases with its temperature.

   a. Consider the following information about the relative strengths of spectral lines due to ions in stars. Rationalize this information using what you know about stellar temperatures (Table 2-1) and ionization energy (Table 3-1).

   • Spectral lines of Si^+ are strongest in Class A stars,

   • Spectral lines of Si^{2+} are strongest in Class B stars, and

   • Spectral lines of Si^{3+} are strongest in Class O stars.

   b. The spectral lines of Ca^+ are strongest in G and K stars, the lines of Si^+ are strongest in A stars, and the lines of He^+ appear in O stars. Does this pattern agree with what you know about stellar temperatures and with the ionization energies in Figure 3-6? Explain your reasoning.

What do stellar spectra tell us? Part II

Many-electron Atoms

Exploration 4A

How are the spectra of atoms like fingerprints?

Creating the Context

Thus far, we've observed spectral lines of hydrogen and of ions such as He$^+$ and Ca$^+$ in stellar spectra. From this we may conclude that these elements are present in stars. However, since stars are the source of all natural known elements, it is likely that stars are composed of other elements as well. How do we find out from its spectrum which other elements are in a star? Our goal for this session is to continue exploring answers to the question, *What can stellar spectra tell us?*

In Exploration 3C, we used ionization energy data to develop a shell model that described the structure of an atom. By knowing how electrons are arranged in an atom, we can better describe how radiation can rearrange these electrons. In Exploration 3D you measured the spectrum of hydrogen and found that it is accurately described by the Bohr model. According to this model, the electron can make a transition from one orbit of *quantized* energy to another only by absorbing or releasing energy that corresponds to the energy difference between the two orbits. You learned that electron transitions via absorption or emission of specific wavelengths of radiation produce the discrete hydrogen spectrum.

Using the spectrum of hydrogen as a fingerprint, you were able to identify hydrogen in a star. Our next task is to use stellar spectra to find other elements in stars. If the spectra of many-electron atoms can also be used as fingerprints, we will be able to use the characteristic spectral lines of these atoms to identify their presence in stellar spectra. In this Exploration we will examine the spectra of several elements and determine, *How are the spectra of atoms like fingerprints?*

The Bohr model worked so well in explaining the spectrum of hydrogen that scientists tried to apply it to other atoms. However, even for helium with only two electrons, it failed to predict the experimentally observed spectrum. It soon became clear that Bohr's theory could not be extended to atoms with more than one electron. Thus, in this Session we will revise our current model of the atom to one that applies to all atoms.

Developing Ideas

Spectra of many-electron atoms

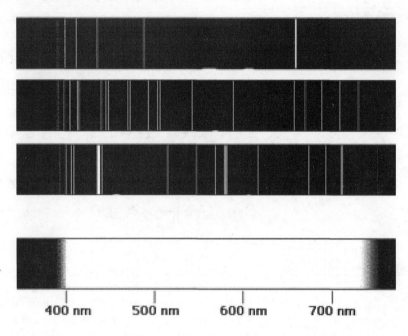

Figure 4-1:   Visible line spectra of hydrogen (top), helium (middle) and mercury (bottom).

1. Examine the spectra of hydrogen (top), helium (middle) and mercury (bottom) in Figure 4-1 above. Helium (He) has two electrons, while mercury (Hg) has 80 electrons. Compare the spectrum of H with one electron to those of He and Hg with many electrons. What do you notice about the spectra of many-electron atoms?

2. Based upon your observations of these three atomic spectra, do you think the spectrum of an atom can be used as its fingerprint? Explain your reasoning.

Exploration 4B

How can we describe the wave nature of electrons?

Creating the Context

In Exploration 4A we learned that the spectra of a many-electron atom can be used as its fingerprint, but that the Bohr model cannot accurately predict such atoms spectral wavelengths. Thus, it evident that we need to improve our model of the atom to include many-electron and one-electron atoms and ions.

Based upon our observations of the properties of light in Session 1, we have determined that light can behave as both a wave and a particle. In 1924, Louis de Broglie proposed a hypothesis that like light, matter such as electrons can have properties of both particle and wave. What implications does a wave nature of electrons have for our understanding of atomic spectra? To answer this question we need to understand the difference between traveling and standing waves.

Thus far you have learned that electromagnetic waves can have any wavelength and frequency, and that the energy of radiation is given in terms of frequency, as E=hν. Electromagnetic waves are an example of **traveling waves**, which move through space with speed *c*. Another type of wave is a **standing wave**, which vibrates in a fixed region of space. Standing waves are produced by a guitar when strings tied down at both ends are made to vibrate, and they are responsible for the music we enjoy. De Broglie suggested that standing waves might be related to the energy states possible for electrons. To understand how, we need to examine de Broglie's hypothesis in greater detail and explore the implications it has for our model of the atom. In this Exploration we will begin by examining the characteristics of standing waves, and answer the question, *How can we describe the wave nature of an electron?*

Preparing for Inquiry

De Broglie and the Wave Properties of Electrons

Although at the time there was no evidence in support of his theory, de Broglie proposed that electrons have particle *and* wave properties. He summarized this hypothesis in the following equation for the wavelength of an electron:

$$\lambda = h / p = h / mv$$

where h is Planck's constant and p is the momentum, equal to the product of mass (m) and velocity (v). In our study of the nature of light in Session 1 we learned that the observation of diffraction and interference is evidence for wave properties of light. Similarly, verification of de Broglie's theory depended upon observations of diffraction and interference of electrons.

These important observations were first made by C. J. Davisson and L. H. Germer in 1927. They found that a beam of electrons directed at a piece of pure nickel resulted in a diffraction pattern. They suggested that electrons diffract from spaces between atoms, in the same way that light diffracts from small openings in a barrier (see Exploration 1C). Subsequent diffraction experiments have further demonstrated the wave behavior of electrons. Figure 4-2 below shows a diffraction pattern produced by electron diffraction from two slits. It is the same pattern of light and dark fringes produced by diffraction of light.

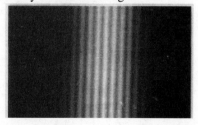

Figure 4-2: The two-slit diffraction pattern produced by a beam of electrons (© Claus Jönsson).

De Broglie's wave theory is thought to apply to all matter (must have mass), to both microscopic and macroscopic objects. It would be hard to prove, however, that a marathon runner behaves as a wave. The mass of a human being is about 10^{32} times greater than an electron, and thus its wavelength would be insignificant compared to the diameter of an atom.

Heisenberg and the Consequences of Wave Properties

In the Bohr model of the atom, we viewed electrons as particles that orbit the nucleus in a circular trajectory. Using well-established ideas from physics, we could determine both the position and momentum of electrons in an atom, just as we could for a baseball thrown into the air. The wave nature of electrons, however, casts doubts upon our ability to know where an electron is and how it is moving. To see why this might be true, we need to think more about how we relate the properties of waves to the location and momentum of particles.

We already know that waves can be characterized by their wavelength and amplitude. Thus, according to de Broglie's equation $\lambda = h / p$, if we can measure the wavelength of a wave associated with a particle like the electron, we will also know the momentum of that particle.

To determine the location of this particle, we need to refer back to electromagnetic waves. According to physics theory, the **probability** or likelihood of detecting a photon at a given point in space is determined by the square of the amplitude of the electromagnetic wave at that point. This implies that the chance of finding a particle at some point is greatest where the amplitude of the wave associated with the particle is largest.

Let's test this by measuring the location and momentum of particles described by several waves. First, consider a particle described by the wave in Figure 4-3. We can easily measure the wavelength of this wave and thus determine the momentum of the particle. Since the amplitude is the same everywhere, however, the particle is equally likely to be anywhere in this region.

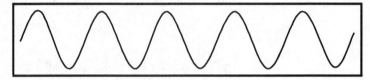

Figure 4-3:   Example of a wave associated with a particle that could be anywhere in the region.

Now consider the waves in Figure 4-4. These are examples of *localized waves*, in which the amplitude is greatest in some small region of space. The wave on the left is more localized in space than the wave on the right, implying that we can be more certain of the location of particle A than particle B. What can we say about the momentum of these particles?

The de Broglie equation (see page 49) tells us that in order to determine the momentum of each particle, we need to know the wavelength. However, because there are few complete waves to measure for particle A, any measurement of the wavelength will involve uncertainty. In contrast, there is less uncertainty in the wavelength of particle B because there are several complete waves that we can measure. As a result, we find that the more localized the wave, the more *certain* we are of its position, but the more *uncertain* we are of its momentum.

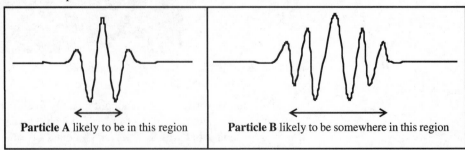

| **Particle A** likely to be in this region | **Particle B** likely to be somewhere in this region |

Figure 4-4:   Examples of localized waves associated with two particles.

In 1927, Werner Heisenberg summarized these observations in a statement about the consequences of the wave nature of matter. The **Heisenberg Uncertainty Principle** states that while you *can* know the location of an electron and how it's moving, you cannot know both things at once without some uncertainty. In other words, there is a limit to how accurately we can simultaneously measure the position and the momentum of an electron. This limit is often expressed by the following equation:

$$(\Delta x)(\Delta p) \geq \frac{h}{4\pi}$$

where Δx and Δp represent uncertainty in position and momentum, respectively, and h is Planck's constant.

Developing Ideas

Electron Diffraction Results

Examine Figure 4-2 of this module, which shows the image produced on a screen when a beam of electrons is sent through two slits in a barrier.

1. What does this experimental result mean in terms of the nature of the electron? Explain your reasoning in terms of what you know about the properties of light.

How do electrons behave like waves?

De Broglie suggested that electron waves might be similar to a special type of wave called a standing wave. In this activity you will have a chance to see how standing waves are produced by a length of rope fixed at both ends, and observe some of their properties.

2. Observe the standing waves and sketch the wave patterns you observe in the space below.

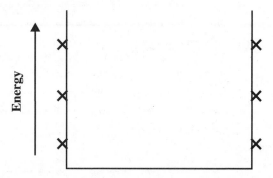

X = position of rope ends

3. What do you notice about the energy (or vibration frequency) required to produce the standing waves? Can a standing wave have energy of *any* value? Explain.

4. The points on a standing wave that have zero amplitude are called **nodes**. Label each standing wave in your sketch with the number of nodes. What relationship, if any, do you notice between the number of nodes and the energy of the standing wave?

5. Consider what you know about the energy of an electron in an atom from hydrogen in Exploration 3D. Can an electron in an atom have *any* amount of

energy? How would you relate the energy quantization of electrons in an atom to standing waves? Briefly explain your reasoning.

Particle-Waves According to Heisenberg

6. How much can we say about where electrons are in an atom if electrons behave as waves? Examine Figure 4-4 of the module:

   a. The wave amplitude squared represents the *probability* of finding an electron along the x-axis. Thus a large wave amplitude will lead to a greater probability of finding the electron in that region of space. Which particle's location do you know better, particle A or B?

   b. Which particle's momentum, A or B, do you know better? (Recall the de Broglie equation $\lambda = h / p$. What do you need to know about the wave to know momentum?)

Applying Your Ideas

7. Use the de Broglie equation to estimate the wavelength of a baseball pitched at 90 miles/hour.

8. Calculate the de Broglie wavelength of an electron in an atom ($m_e = 9.11 \times 10^{-31}$ kg) moving with the speed of 10^6 m / s. Calculate your wavelength either running or walking. (Estimate your velocity knowing that the average person runs about 5 miles / hr and walks about 2 miles / hr.)

9. A criterion for testing the wave properties of an object is to compare its de Broglie wavelength to the region of space where it is confined. If these values are comparable, this means that the wave-nature of the object is important.

   a. For an electron in an atom, this region would be determined by the diameter of the atom (on the order of Angstroms).Use this criterion to describe the extent of the wave properties of electron.

   b. We typically live **in** regions of space on the order of meters. Use the criterion above to describe the extent of your wave nature. Whose wavelength is more significant to its behavior, yours or the electron's?

10. Does the Heisenberg Uncertainty Principle state that you can *never* know the position and the momentum of an electron simultaneously? Explain your reasoning.

Exploration 4C

What wave model explains the structure and spectra of atoms?

Creating the Context

You learned in Exploration 4B that a beam of electrons directed at a double slit will produce interference and a diffraction pattern. This evidence of wave behavior confirmed de Broglie's hypothesis that electrons have both particle and wave properties. As a consequence, you learned that we can't simultaneously know the position and momentum of an electron without some uncertainty in our knowledge. This result, called the Heisenberg Uncertainty Principle, indicates that we need to develop a new language for talking about the location of electrons. In this Exploration we will improve our Bohr model of the atom to account for the wave properties of electrons and answer the question, *What mathematical model describes the structure of atoms?*

As a first step in developing a model for the atom that describes its wave nature, we considered the properties of standing waves. In Exploration 4B we saw that standing waves are produced only for specific vibration frequencies or energies. We also learned that each standing wave of different energy can be characterized by its number of nodes, or points on the wave of zero amplitude. The number of nodes directly corresponds to the energy of the wave, so that the lowest energy standing wave has zero nodes, the next higher energy wave has one node, and so on. These observations demonstrate that standing waves lead to a *quantization* of energy of the wave. Thus, a general model of the atom that describes the wave nature of matter should incorporate the idea of standing waves.

Preparing for Inquiry

Quantum Model of the Atom

De Broglie's theory and electron diffraction experiments illustrate the need for a model that incorporates the wave nature of matter, predicts the same results for the spectrum of hydrogen as the Bohr model, and explains the spectra of many-electron atoms. Such a model is called **quantum mechanics**. Quantum mechanics combines the characteristics of both particles and waves by proposing that a particle like an electron can also be represented as a wave. The possible energies and locations available to an electron are described by a mathematical function called the **wavefunction**, symbolized by the Greek letter, Ψ ("psi").

A wavefunction is a mathematical expression that relates the movement of a wave in space to the amplitude of the wave at any point in space. For example, the standing wave with *one* node you observed in Exploration 4B can be described by a wavefunction $f(x) = \sin(x)$, where $f(x)$ is the amplitude of the wave at a particular point in space. Similarly, the wavefunction $\Psi(x,y,z)$ tells you the amplitude of the wave associated with a particle at a point (x,y,z) in space. According to this quantum model, the key to understanding how an electron is arranged in an atom depends upon knowing the values of the function $\Psi(x,y,z)$. But where do wavefunctions come from?

In 1925 Erwin Schrödinger wrote down an equation called the **Schrödinger equation.** The solutions to this equation give two types of answers, wavefunctions and the quantized energies of these wavefunctions. The wavefunctions correspond to the quantized stationary states or "orbits" of an atom first introduced by Bohr (see Exploration 3D). Although we call these wavefunctions **atomic orbitals**, unlike Bohr orbits, they *do not* describe the trajectory or path of an electron.

What *does* the wavefunction associated with an electron tell us about its location? We know from the Heisenberg Uncertainty Principle that we can't regard the electron as a localized charge orbiting the nucleus. Thus, we need an alternative way of thinking about it. Since the wave associated with an electron is spread out in space, one possibility is to think of the electron as a cloud, with charge spread out about the nucleus. This charge, however, is not equally distributed; some regions have more charge than others, and these regions correspond to a higher probability of our finding the electron.

We learned in Exploration 4B that the amplitude of an electromagnetic wave at a point in space is related to the probability of finding a photon at that point. More precisely, the probability is equal to the square of the amplitude, A^2. By analogy, quantum mechanics suggests that the square of the amplitude, $\Psi(x,y,z)^2$, is the probability of finding a particle at the point (x,y,z). We often plot the **electron probability density**, Ψ^2, as a way to visually represent atomic wavefunctions, Ψ, in three dimensions. Chemists use electron probability density plots to represent the wavefunctions of atoms so frequently that when we use the term *atomic orbital*, we often picture an electron probability plot. In this Exploration we will learn more about the shapes and descriptions of orbitals, and how we can use them to describe where electrons are most likely to be in an atom.

Orbitals of Hydrogen

The orbitals, Ψ(x,y, z), of hydrogen used to describe the location of its electron are often identified with **quantum numbers** that appear in the wavefunctions. The quantum number n describes the *size* and *energy* of an orbital and is related to the number of nodes. As the value of n increases, so does the energy of the electron. Another quantum number, symbolized by the letter l, describes the *shape* or type of orbital. In three dimensions, an orbital of a given type can be oriented in several ways. Thus, we need a third quantum number m (often called m_l) that describes the *direction* of an orbital (See Table 4.1).

According to the shell model in Exploration 3B, the integer n labels shells and the letters s, p, and d label subshells. In quantum mechanics, n is analogous to the shell number. Likewise, the subshells within a shell correlate to different values of **l.** l = 0 is an **s orbital**, l = 1 is a **p orbital**, l = 2 is a **d orbital**, l = 3 is an **f orbital**, and so on. We refer to the orbitals with the same n quantum number as a shell. For example, the valence shell in carbon includes the orbitals with n = 2, the 2s and 2p orbitals.

The limits upon the numerical values of n ,l, and m are determined by solving the Schrödinger equation. We can use these quantum numbers to describe the location of the electron in hydrogen in terms of the orbital $\Psi_{n\,l\,m}$ it occupies.

Table 4-1

Quantum Number	Description	Limits
n	size and energy of orbital	integer, n = 1 or greater
l	shape of orbital	integer, l = 0 to (n - 1)
m	orientation of orbital	integer, m = 1 to -l, including 0

Many-electron Atoms

As the number of electrons increases in an atom, solving the Schrödinger equation to determine wavefunctions becomes very complicated. Fortunately, if we make certain approximations about the way electrons interact with each other and the nucleus, we can use hydrogen-like orbitals ($\Psi_{n\,l\,m}$) to describe the location of electrons.

The principle that governs the electron structure of atoms is called the **Pauli Exclusion Principle**. It says that each atomic orbital, $\Psi_{n\,l\,m}$, can accommodate at most two electrons. It also states that no two electrons can have the same set of quantum numbers. However, if two electrons are in the same orbital, they *should* have the same quantum numbers! We can resolve this apparent contradiction by introducing **m_s**, the **spin quantum number**, which can have two values, chosen by convention to be +1/2 ("spin-up") or -1/2 ("spin-down"). Thus, if two electrons occupy a single orbital, one will have m_s = +1/2, and the other m_s = -1/2.

Developing Ideas

Wavefunctions in One Dimension

You saw in Exploration 4B that standing waves in one dimension (1D) have nodes, and that the number of nodes in a given wave is proportional to its energy. In 1D, nodes are points. Wavefunctions $\Psi(x)$ associated with particles constrained to movement in one dimension are the one dimensional standing waves.

1. First consider the two wavefunctions sketched below. A particle described by a 1D wave is restricted to moving in one direction.

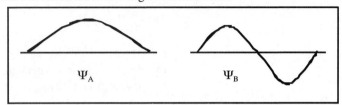

Ψ_A Ψ_B

   a. Sketch the approximate shape of the probability density, Ψ^2, for Ψ_A and Ψ_B above. Do the location and number of nodes change? *Hint: Think about what happens to the sign and magnitude of a negative or positive amplitude when you square it.*

   b. For each Ψ^2 sketch in part a, indicate with an arrow the region(s) where a particle described by Ψ would most likely be located.

Orbitals of Hydrogen

The three-dimensional wavefunctions, Ψ, for an electron in an atom are so important that they are given the name **atomic orbitals**. It is easier to visualize atomic orbitals by plotting electron probability density, Ψ^2. Examples of these electron probability plots can be found in your textbook. They resemble clouds, where the denser regions of the cloud indicate a higher probability of finding the electron. Atomic orbitals have characteristic shape, size, and energy.

2. Use your textbook to draw one example of an *s*, *p*, and *d* orbital.

   a. Describe the characteristic shape of each of these orbitals.

   b. Orbitals represent the probability of finding an electron in a particular region of space. In this framework, what does a node mean?

   c. Which type(s) of atomic orbitals have nodes, or regions where there is zero probability of finding the electron? Which do not have nodes? Identify any nodes in your orbital sketches above.

3. The size and energy of an orbital depends on the n number that describes it. For example, the 1s, 2s, and 3s atomic orbitals have different sizes and energies.

   a. Use your textbook to sketch and label the 1s, 2s and 3s orbitals, paying attention to relative size.

   b. Atomic orbitals also give you information about the chance of finding an electron in an atom at some distance from the nucleus. We imagine the nucleus at the radial center of an orbital. Predict which electron will be found *furthest* from the nucleus, a 1s, 2s or 3s electron. (The n=1,2,3 value has similar meaning to shell number n under the shell model.) Explain your reasoning.

   c. According to Coulomb's law, which electron, 1s, 2s, or 3s, will require the *least* energy to be removed from an atom?

   d. Use your understanding of the hydrogen atom to speculate which of these orbitals has the highest energy. Which orbital has the lowest energy? Justify your answers.

Naming Orbitals

4. Your instructor may show you a video about atomic orbitals to help you visualize where electrons are in an atom. Write at least two of your observations.

5. Name the orbitals of hydrogen by completing the chart of possible quantum numbers below (see Table 4-1). We often use the notation "n𝑙" to name an orbital. For orbitals with 𝑙≠ 0, we add a subscript for the different orbital orientations, or values of m. By convention, we use the subscripts x, y, and z, for the 𝑙= 1 orbitals (e.g. 2p$_x$), and integer subscripts 1, 2, 3, etc. for 𝑙> 1.

n	𝑙	m	orbital (Ψ_{nlm})	orbital name
1	0	0	Ψ_{100}	1s
2				
2				
2				

Applying Your Ideas

6. Orbitals describe where you are most likely to find an electron. The radial centers of these orbitals represent where the nucleus is located. Consider the *p* orbital with one node. How likely is it that you will find a *p* electron at the nucleus of an atom? Explain your reasoning.

7. Where (in what orbital) is the electron of hydrogen in its lowest energy (ground) state?

8. Which set(s) of quantum numbers for hydrogen is <u>not</u> allowed? For any set that is incorrect, state why *and* write one correct set of numbers.

    a. $n = 2, l = 0, m = 0$

    b. $n = 3, l = 3, m = -2$

    c. $n = 5, l = 2, m = 4$

Many-Electron Atoms

All orbitals of hydrogen with a given value of n have the same energy. However, in the presence of interactions between electrons, orbitals of many-electron atoms with the same n *do not* have the same energy. Thus, to understand where the electrons are in many-electron atoms, we need to take into account the energy order of the orbitals.

Recall that orbitals are filled from lowest to highest energy (see page 36), one electron at a time. When an orbital is "filled" by an electron, this means that the electron is described by that orbital, and hence the energy of the electron is the energy of the orbital. As required by the Pauli Exclusion Principle, there can be *at maximum* two electrons with opposite spin in each orbital. An additional set of guidelines, called **Hund's Rules**, can help you decide on the lowest energy, most favorable electron configuration. These guidelines are

a. When filling orbitals of the *same* energy, first place one electron in each of these orbitals, and then place a second electron in any of the orbitals. (e.g. the following three different p orbitals have the same energy)

b. If electrons in different orbitals are unpaired, placing these unpaired electrons either all spin-up or all spin-down will give the lowest energy configuration. (See the example for nitrogen in Figure 4-5.)

In Exploration 3C you learned to write electron configurations using a $1s^2 2s^2$ type of notation. We can also write electron configurations using **box diagrams**, which symbolize each orbital with a box and each electron with an arrow pointing either up (spin-up) or down (spin-down), and fill each orbital according to Hund's rules and the Pauli Exclusion Principle. For example, we would write the electron configuration of lithium and nitrogen in box notation as shown in Figure 4-5.

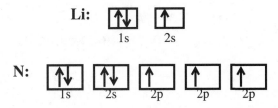

Figure 4-5: Electron configurations of lithium and nitrogen using box diagram notation.

9. Use box notation to write the electron configurations of the following elements:

    a. C

    b. O

  c. Na

  d. Fe

10. Write the set of quantum numbers (n, l, m, m_s) that describes *one* of the 2p electrons in carbon.

11. How many electrons can have these sets of quantum numbers?

  a. n =3, l =2, m $\ddagger$, m_s= +1/2

  b. n = 2

  c. n = 3, l = 2

Exploration 4D

How can the spectra of many-electron atoms tell us what is in a star?

Creating the Context

Thus far in this module you have learned much about the structure of stars and the properties of the light they emit. In addition, you have discovered that the interaction of radiation with an atom yields information about its electronic structure. Much of the information we have today about the structure of atoms can be attributed to studies of atomic spectra.

In Exploration 3D you used the Bohr model to predict the spectrum of hydrogen, and the spectrum to identify the presence of hydrogen in a star. In Exploration 4A, however, you saw that the spectra of many-electron atoms are more complicated than hydrogen and cannot be described by the Bohr model. In Exploration 4B, you learned that electrons have both wave and particle properties, in much the same way as light does. Thus, we proposed a revised model of the atom called *quantum mechanics* to explain the structure and spectra of many-electron atoms. Now, in order to fully characterize the composition of a star, we need to use our revised model of atomic structure and determine, *How can the spectra of atoms tell us what is in a star?*

To begin, you will observe the energies of visible light emitted by fluorescent lights and the light from the sun, and use spectroscopes to observe the spectrum of each light source. You will also examine the spectra of two stars, and use atomic spectral wavelengths to analyze these spectra.

Developing Ideas

PART I: YOUR EYES AS DETECTORS

Use your eyes to observe the colors of the following sources of light. Record these colors below. **NOTE:** Do not look directly at the Sun as serious eye damage may result.

1. The Sun:

2. Fluorescent lights:

3. A variety of flames: (see instructions below)

When various solids are placed in a flame, the solids are vaporized into the component atoms. The atoms that absorb heat energy from the flame are referred to as excited. The light energy emitted by these excited atoms can be detected as the color of the flame.

• Insert the tip of the flame test wire into a solid or solution of the element. Then, place the metal tip into a bunsen burner flame, and observe the color. Repeat this for each element.

Element	Color

PART II: A SPECTROSCOPE AS A DETECTOR

Flames

4. Observe the emission line spectra of the various atoms used in the flame tests above. Use the **Elements** web tool found on the Module Web Tools CD-ROM in Exploration 4D of the Stars web site. Alternatively, your instructor will provide the internet address for this page to you.

5. Describe how the color(s) of emission lines in the line spectrum of each element compare with its flame color.

Fluorescent light

The illumination from a fluorescent light is obtained by filling a glass tube that contains metal electrodes at each end with a gas at low pressure. When a high voltage is placed across the electrodes, a stream of fast-moving electrons travels through the gas from one electrode to the other. Some energy is transferred to the atoms that make up the gas. The atoms that have received this energy are referred to as excited. One way the excited atoms release their energy is as light.

6. View a fluorescent light through a spectroscope.

7. Record the wavelength(s) of the light energy emitted by the fluorescent light in the table below. Also record the color associated with each visible wavelength of light.

Wavelength (nm)	Color	Element

8. Characteristic wavelengths for various elements are listed in Table 4-2. By comparing the wavelengths for the emission lines that you have observed with those in the table, identify which element(s) are emitting light in the fluorescent light.

9. How might you reconcile the color of the fluorescent light you saw with your eyes with the observation of its spectrum you made with the spectroscope?

Sunlight

When you look at the light from our Sun through the spectrometer, you will see a rainbow of radiation with a few dark lines superimposed. The dark lines are caused by elements in the outer atmosphere of the Sun that absorb certain wavelengths of the light. Because each element absorbs light of characteristic wavelength, by measuring the wavelengths of the dark lines, you can identify the elements in the Sun that absorbed the sunlight.

 Caution Serious eye damage can result from pointing an optical device at the Sun when your eye is in the path. Lenses and mirrors focus sunlight and can set paper on fire as well as damage your eye.

10. View sunlight through a spectroscope.

11. Record the wavelengths of the dark lines in the solar spectrum in the table below.

Wavelength (nm)	Element

12. Characteristic wavelengths for various elements are listed in Table 4-2. By comparing the wavelengths for the dark lines that you have observed with those in the table, identify which element(s) are absorbing light from the Sun (causing dark lines) and are therefore present in the atmosphere surrounding the Sun.

The lines you see were first noted by Josef von Fraunhofer in 1813. At a young age von Fraunhofer was apprenticed to a Bavarian glazier and thus worked with glass and light. He later produced a very sophisticated diffraction grating and used the grating to record the spectra of the Sun and stars. Using this spectrometer, he found that the Sun and stars have different spectral lines. This was the basis for the first proof that stars make their own light, while the planets merely reflect the Sun's light.

PART III: ANALYSIS OF STELLAR SPECTRA

In Exploration 2A you learned of a spectral classification scheme in which stars are grouped into 7 classes: O, B, A, F, G, K, M. Each of these classes can be further divided into sub-classes numbered 0-9, from high (0) to low (9) temperature. In this activity, you will apply what you have learned about the spectra of atoms to determining the elements present in the spectra of two stars.

13. Measure the wavelengths of the dominant peaks in *each* of the stellar spectra (F3 and B5) in Figure 4-6 below. The dominant peaks are indicated with arrows.

14. Use Table 4-2 to assign these spectral wavelengths to various elements. Record your assignments beneath the arrows in the spectra.

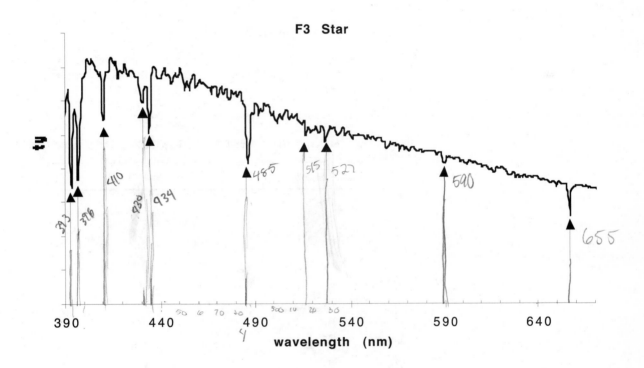

F3 Star

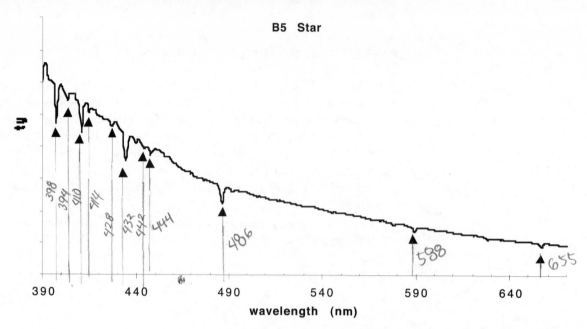

Figure 4-6: Spectra of a F3 and B5 star, from the holdings of the Astronomical Data Center (ADC), Silva, D., "A New Library of Stellar Spectra" (1992).

Table 4-2   Common Spectral Lines for Some Elements

λ(nm)	Element	λ (nm)	Element
378	Ne	461	Sr
393.5	Ca^+	468.6	He^+
396.8	Ca^+	476.1	TiO (molecule)
397	H	478	MgH (molecule)
400.5	Fe	486	H
403	He	492	Fe
405	Hg	517	Mg
407.8	Sr^+	518	Mg
410	H	527	Fe
413	Li	541.2	He^+
414.1	Fe	546	Hg
420	He^+	554	Ba
421.6	Sr^+	577	Hg
422.6	Ca	579	Hg
425	Fe	585	Ne
427.1	Fe	589	Na
430	CH (molecule)	590	Na
431	Fe	610	Li
434	H	616.2	Ca

Table 4-2   Common Spectral Lines for Some Elements

λ(nm)	Element	λ (nm)	Element
436	Hg	618	Ne
438.4	Fe	630	Fe
439	Ne	641	Sr
440	Ti^+	650	Fe
440.5	Fe	656	H
445.5	Ca	671	Li
447.1	He	696	Ar
454.2	He^+	707	Ar

Applying Your Ideas

The following problems will give you more practice with the concepts.

15. **Stellar Spectrum.** Summarize the elements that are present in the stellar spectra you analyzed and the process you used to determine these elements.

16. **The solar spectrum.**   In 1811, Joseph von Fraunhofer studied sunlight dispersed into its component colors, as you did with the spectroscope. The solar spectrum is shown below. Although he expected to see visible light of all energies, von Fraunhofer found that narrow bands of energy were missing, causing the dark bands shown here. What process causes these dark bands? *Hint: Recall the structure of a star discussed in Exploration 3A. What can happens to photons in the atmosphere of a star?*

We learned in Exploration 3D that only specific wavelengths of radiation can be absorbed or emitted by hydrogen. On the atomic level, you learned that absorption and emission correspond to electrons making transitions between different energy levels or shells. Now we have refined our model of the atom; we talk about orbitals and regions of electron probability rather than shells. However, absorption and emission have similar meanings under the orbital model of atoms; electrons simply move between orbitals of specific energy.

17. **Neon signs.** The neon signs we see on storefronts are colored because of atomic emission. When neon atoms are bombarded with electrons in a discharge tube, emission of orange-red light is observed. Although the spectrum of neon consists of many discrete wavelengths, our eyes perceive a mixture of these wavelengths. In terms of what you know from Exploration 4C about the orbital model of atoms, explain why only discrete wavelengths of emitted light are observed in neon.

18. **Sodium lamps.** Many parking lots are lit up with the yellow glow of sodium lamps at night. These lamps are discharge tubes filled with gaseous sodium atoms. Based upon your knowledge of the structure of atoms from Exploration 4C, do you think that the yellow light emitted by sodium atoms in the lamps contains *every* wavelength of yellow (approximately 525 to 575 nm) in the visible region of the electromagnetic spectrum? Explain why or why not.

19. **Fluorescent lights.** Fluorescence is an emission process similar to the emission of light by hydrogen. However, in the case of fluorescence, atoms emit light of *longer* wavelengths than the wavelengths of light initially absorbed. A fluorescent lamp is a discharge tube whose inner surface is coated with a fluorescent material such as zinc sulfide. The tube is filled with mercury vapor at low pressure. Electron bombardment of the mercury atoms causes the emission of light

in the green, blue, and ultraviolet (UV) regions as you observed with the spectroscope. When the UV light strikes the inner wall, it is absorbed by the fluorescent material, which then emits light of many *longer* wavelengths to produce the white light that we see.

a. Zinc sulfide is colorless. What does this tell you about the ability of zinc sulfide to absorb white (visible) light?

b. Explain why ultraviolet light causes the zinc sulfide to fluoresce, but visible light does not. Hint: If zinc sulfide could absorb white (visible) light would the resulting fluorescent radiation be visible to us?

Exploration 4E

How does the wave model explain the properties of atoms?

Creating the Context

An essential test of the validity of any new theory is whether it can predict experimental results. Our current model of the atom called quantum mechanics accounts for the wave nature of electrons and the discrete nature of atomic spectra. An additional test of the model is to use it to explain the periodic trends in atomic properties we first encountered in Exploration 3B. Thus, the goal of this Exploration is to respond to the question: *How does the wave model explain the properties of atoms?*

Some familiar properties of atoms are ionization energy and atomic radii. If our wave model of the atom is valid, we should be able to use it to explain the experimentally observed trends in these properties. We know that the energy of an electron in an atom reflects how strongly it is attracted to the nucleus. This attraction determines how much energy will be required to remove the electron from the pull of the nucleus (ionization energy). The energies of electrons are also determined by their distance from the nucleus. In a given atom, electrons that are close to the nucleus are strongly attracted to it, while those that are farther from the nucleus are weakly attracted to it. Thus, to test the validity of our model, we will need to know more about the energies of electrons in an atom.

You already know that according to Coulomb's law, the energy of attraction between an electron and the nucleus is given by their charge and the distance between them. In hydrogen, we need only consider the interaction between a single electron and the nucleus. However, in a many-electron atom, there are electron-electron interactions as well as electron-nucleus interactions. Given this complexity, how do we think about the energy of electrons in many-electron atoms? In this Exploration, we will examine one approximation technique to the wave model of atoms that simplifies these multiple interactions. Under this approximation, we can use the wave model to describe the energies of electrons in many-electron atoms, and also to explain trends in the properties of atoms.

Preparing for Inquiry

A shielding approximation to the wave model

As a start, consider the elements Li, B, and Fe, with the electron configurations:

- Li: $1s^2 2s^1$
- B: $1s^2 2s^2 2p^1$
- Fe: $1s^2 2s^2 2p^6 3s^2 3p^6 4s^2 3d^6$

These electron configurations imply a certain energy order of the orbitals (and the electrons that are described by these orbitals) of many-electron atoms. Specifically, the filling orders for Li and B imply that 1s is lower in energy than 2s, and that 2p is lower in energy than 2p. Why is this the case? The electron configuration of Fe is also puzzling, since it indicates that the 4s orbital is lower in energy than the 3d orbital. How can we think about the energy of electrons in many-electron atoms?

An electron in a many-electron atom interacts with other electrons and the nucleus according to Coulomb's law,

$$E_{attraction} = -kZe^2/r.$$

It experiences Coulombic repulsions from all other electrons and attraction to the nucleus. Thus, to understand the energy of an electron in a many-electron atom, we need to think about the net outcome of these repulsive and attractive interactions. Consider helium, with two electrons and nuclear charge 2e. If the electrons did not repel each other, each electron would "see" a nuclear charge of 2e. However, electrons *do* interact and repel each other. How does the electron-electron repulsion affect the nuclear charge that each electron "sees"?

Let's examine the behavior of one electron of helium in the presence of the other. Figure 4-7 shows both electrons as spherical electron clouds. If electron 2 is far away from the nucleus and from electron 1 ($r_2 > r_1$), then electron 2 will be repelled by the part of the electron 1 cloud that is within the r_2 distance. As a result of this interaction, electron 2 will be less attracted to the nucleus and experience a reduced nuclear charge, or an **effective nuclear number**, Z_{eff}, where $Z_{eff} < Z$. We say that electron 2 is **shielded** from the nucleus by electron 1. In the limit that r_2 is very large, electron 2 experiences almost complete shielding, and Z_{eff} approaches unity.

If, however, electron 2 is close to the nucleus so that r_2 is within the r_1 distance ($r_2 < r_1$), then electron 2 is *less* shielded by electron 1, and so electron 2 feels a *greater* effective nuclear number, Z_{eff}. In this case, Z_{eff} is closer to the true atomic number, $Z = 2$. In the limit that r_2 becomes very small, electron 2 experiences virtually no shielding, and Zeff approaches the unshielded nuclear number, $Z = 2$.

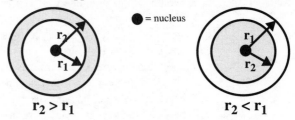

$r_2 > r_1$ $r_2 < r_1$

Figure 4-7: Two relations of the position of electron 2 (r_2) to electron 1 (r_1) in a two-electron atom.

We can see that the effective nuclear number and the amount of shielding from the nucleus experienced by an electron depends on its distance from the nucleus in relationship to the distances from other electrons. In helium, each 1s electron is, on average, the same distance from the nucleus, and the electrons do not effectively shield each other. Hence, we expect the Z_{eff} for each helium 1s electron to be the same, and closer to $Z = 2$; the calculated value of Z_{eff} is 1.69. Using a **shielding approximation** to the wave model, we now have a simplified model of a many-electron atom, with one electron influenced by an effective nuclear number.

Radial Wavefunctions

You now know that the degree of shielding, and hence the Z_{eff}, that an electron experiences depends on its distance and the distances of other electrons from the nucleus. The next question is: How do we determine Z_{eff} and the most probable distance from the nucleus of an electron?

One approach is to consider the orbital or wavefunction, $\Psi_{n\,l\,m}$ that describes the electron. These electron wavefunctions can also be written as the product of two functions, a radial function and an angular function. The radial function, R_{nl}, depends on distance r and the quantum numbers n and l and it tells you how the amplitude of the wavefunction changes as a function of distance from the nucleus. The angular function tells you how the amplitude of the wavefunction changes as a function of orientation in space. Figure 4-8 below shows a plot of the radial wave-functions, R_{nl}, of 1s, 2s, and 2p electrons.

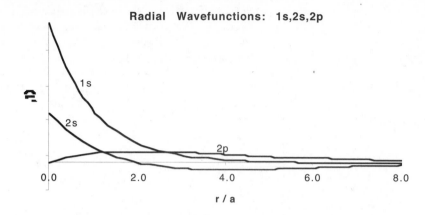

Figure 4-8: The radial wavefunctions for 1s, 2, and 2p electrons.

Since we would like to know the most probable distance from the nucleus of an electron, it is useful to plot the square of the radial wavefunction, R^2, which is related to the orbital probability density, Ψ^2 (see Exploration 4C). A plot of R^2 gives the electron density as a function of distance, r, along a particular direction in space. We can also determine the radial probability in *all* directions by multiplying R^2 by the distance squared, r^2. The plot of r^2R^2 shown in Figure 4-9 below represents the **radial probability distribution** for a 1s electron. Here, the function r^2R^2 on the y-axis is labeled "Probability."

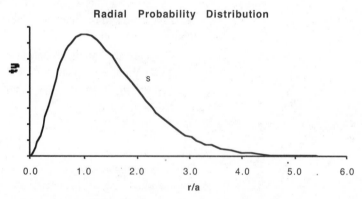

Figure 4-9: Radial probability distribution plot for a 1s electron.

Both the radial function, R, and the radial probability, r^2R^2, are most easily plotted in terms of a dimensionless distance variable, r/a, where r is distance in Angstroms, and a is the **Bohr radius** and has the value 0.529 Angstroms. The Bohr radius first appeared in the Bohr model as the radius of the lowest energy orbit. However, the Bohr radius is also the most probable distance from the nucleus at which to find a 1s electron. Figure 4-9 shows that r/a = 1.0 corresponds to the maximum probability, and thus the most probable distance is r = a = 0.529 Angstroms.

In Exploration 3C you learned how to "fill" atomic orbitals with electrons according to the Aufbau Principle, and to write electron configurations for various atoms. The Aufbau Principle says that the lowest energy orbital is always filled first, according to the filling order on page 37. What is the origin of this energy order? In this Exploration, we will use ideas of shielding and the wave model from Exploration 4C to answer that question.

Developing Ideas

How can we explain the energies of orbitals in many-electron atoms?

To describe the energy of an electron, we will consider a Coulombic interaction between the electron and an **effective nuclear charge**, $Z_{eff}e$. Information about effective nuclear number Z_{eff} and electron distance from the nucleus can be found in radial probability distributions. Work with your neighbor or in small groups to answer the following questions.

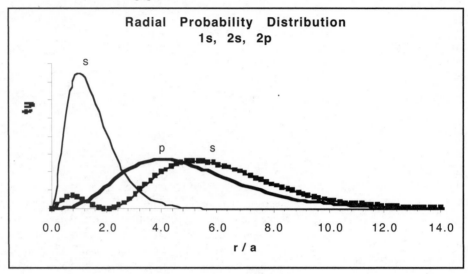

Figure 4-10: Radial probability distribution plots of 1s, 2s, and 3s electron orbitals.

1. The radial probability curves in Figure 4-10 give information about the probable distances from the nucleus of an electron and tell you something about the relative sizes of orbitals.

    a. Compare the radial probability curves for 1s, 2s, and 2p electrons in an atom. Which electron is *most* likely to be found closest to the nucleus? Which curves overlap most, and what might this overlap tell you about size?

    b. How does the variation in size between orbitals in *different* shells compare with variation in size between orbitals in the *same* shell? Explain.

2. The 2s radial probability curve in Figure 4-10 has more than one peak, meaning that although the 2s electron is most likely to be found away from the nucleus, there is a small but finite chance that this electron will penetrate close to the nucleus. This penetration alters the degree of shielding experienced by the 2s electron.

    a. How will the penetration of the 2s electron affect how well it is shielded from the nucleus by inner shell (n = 1) electrons? Explain.

b. An electron that is highly shielded from the nucleus will experience less of the full nuclear charge and have a smaller Z_{eff}. Rank, in decreasing order, the Z_{eff} experienced by 1s, 2s, and 2p electrons in an atom.

3. Use the information about distance from the nucleus and Z_{eff} to rank the energies of 1s, 2s, and 2p orbitals. Does this explain the electron configurations of Li and B (see page 65)? Explain your reasoning.

4. Now, using the radial probability curves in Figure 4-11 below, we will consider the energies of orbitals in the n = 3 shell of an atom.

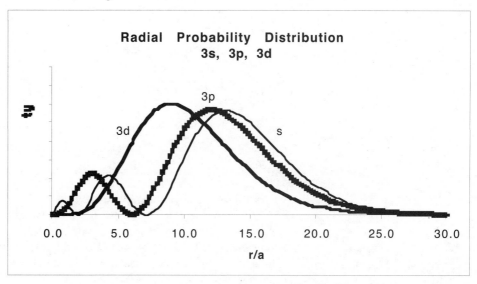

Figure 4-11: Radial probability distribution plots of 3s, 3p, and 3d electron orbitals.

a. Of the 3s, 3p, and 3d electrons, which electron penetrates closest to the nucleus and experiences the *least* amount of shielding by inner shell electrons? Rank, in decreasing order, the Z_{eff} experienced by 3s, 3p, and 3d electrons in an atom.

b. Use information about distance from the nucleus and Z_{eff} to rank the energies of the 3s, 3p, and 3d orbitals.

c. Do your rankings agree with what you know about the filling order of orbitals from the Aufbau Principle and the electron configuration of Fe? Explain your reasoning.

How effectively do different types of electrons shield each other?

You have learned that the same type of electrons (such as two 1s electrons of helium) experience the same degree of shielding from the nucleus (see page 66). In this activity you will explore how effectively electrons in *different* shells and electrons in the *same* shell shield each other.

5. Table 4-3 below shows calculated Z_{eff} values for 1s, 2s, and 2p electrons of C ($Z = 6$). The 2s and 2p electrons are shielded from the nucleus by the inner shell electrons (1s). The 2p electron is also shielded by the 2s electrons in the same shell. Compare the shielding effect by electrons in *different* shells to the shielding by electrons in the *same* shell. Which is more effective? Explain.

Table 4-3

Electron	Z_{eff}
1s	5.67
2s	3.22
2p	3.14

Calculated Z_{eff} values. Source: E. Clementi and D. L. Raimondi, J Chem Phys. 1963, **38**, 2886.

6. Under a shielding approximation to the wave model, Z_{eff} values for the highest-energy valence electron can be calculated for various elements. Use what you know about shielding by electrons of different types (Problem 5) to explain the trends in Z_{eff} in Table 4-4 between He and Li (1s -> 2s), and Li and Be (2s -> 2s). You will also need to consider how the atomic number, Z, changes between elements.

Table 4-4

Element	Z_{eff}
H	1
He	1.69
Li	1.28
Be	1.91

Calculated Zeff values for valence electrons. Source: E. Clementi and D. L. Raimondi, J Chem Phys. 1963, **38**, 2886.

Applying Your Ideas

Properties of Atoms

7. The radial probability distributions of a 3d and 4s orbital are shown in Figure 4-13. Use these curves to explain why Fe has the electron configuration $1s^2 2s^2 2p^6 3s^2 3p^6 4s^2 3d^6$, rather than $1s^2 2s^2 2p^6 3s^2 3p^6 3d^8$.

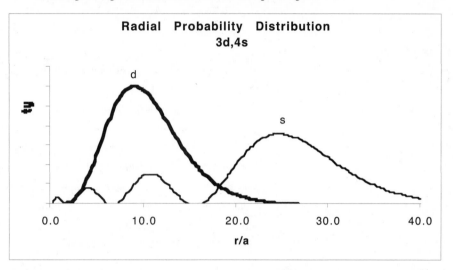

Radial Probability Distribution
3d,4s

Figure 4-12: Radial probability distribution plots of 3d and 4s electron orbitals.

8. Re-examine the plot of first IE in Table 3-6. Does the shielding approximation to the wave model explain the periodic trend in IE between Li and Ne? Explain your reasoning. (Hint: Think about the variation in orbital size between shells and within a shell. Also consider shielding by electrons in the same or different shells. How will these factors affect the energy required to remove the highest-energy valence shell electron?)

9. Use the wave model to explain the trends in atomic radii listed in Table 3-2 down a column, from F to I, and across a row of the periodic table, from Li to C. Use what you have learned about the size of orbitals, and about shielding by electrons in the same or different shells.

Properties of Elements

10. Recall that the ionization energy is the energy required to remove an electron from a *gaseous* element. Metals are characterized by low first ionization energies.

    a. According to this criterion, what region of the periodic table has elements with the most metallic character?

    b. Look back at the work functions (energy required to remove electrons from a *solid*) of various elements shown in Figure 1-10. In general, how do the work functions of metallic elements compare to the work functions of other elements?

11. Which elements in the periodic table have the *least* metallic character, that is, are non-metals? What criterion did you use to make your selection?

12. Many metals are found in nature in ionic form. Do you expect metal ions to be cations or anions? Explain your reasoning.

Making the Link

Looking Back: What have you learned?

Chemical Principles

During Session 4 you learned important chemical principles and terms associated with the structure and spectra of atoms with many electrons. You should be familiar with the following principles:

- Spectra of many electron atoms (4A)
- Standing waves and nodes (4B)
- Electron diffraction and the wave nature of electrons (4B)
- Probability and the Heisenberg Uncertainty Principle (4B)
- Wavefunction, orbital, and electron probability density (4C)
- Quantum numbers, Pauli Exclusion Principle, and Hund's rules (4C)
- Shielding approximation, effective nuclear number, effective nuclear charge, and radial probability distribution (4D)

Thinking Skills

As you worked through Session 4, you developed some general problem-solving and scientific thinking skills that are not specific to chemistry. These skills are valued by employers in a wide range of professions and in academia. You have developed the following skills:

- Evaluating and revising a model (4A, 4C, and 4E)
- Data analysis skills (4C and 4D)
- Data observation and interpretation (4A, 4B, and 4E)
- Interpreting and using mathematical formulas (4B and 4C
- Experimental design (4D)

Checking Your Progress

The following problems will help you integrate the chemistry concepts you have learned in Session 4, connect your understanding to the story line, and begin to make progress toward the culminating activity.

1. In this Session you have learned about the spectra and structure of many-electron atoms. Now, we will use this information to analyze the spectrum of a class G star. The G stellar spectrum in Figure 4-13 below is the same as the one you examined in Session 2, on page 25. There you estimated color and temperature of the star based on λ_{max}. Now, consider the fine features of the spectrum.
   - Assign the spectral lines in the spectrum using your knowledge of the hydrogen spectrum from Exploration 3D (page 38), Table 3-6 (page 46), and Table 4-2 (page 62). Record your results in Table 4-5 below.

G Star

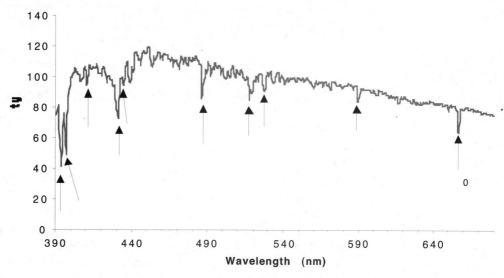

Figure 4-13: Spectra of a G star, from the holdings of the Astronomical Data Center (ADC), Silva, D., "A New Library of Stellar Spectra" (1992).

Table 4-5

Spectral line	Wavelength (nm)	Assignment
1	393.5	
2	397	
3	410	
4	430	
5	434	
6	486	
7	517	
8	527	
9	589	
10	656.5	

2. Write the electron configuration using box notation for each element you have identified in this star.

3. Recall the spectrum of an incandescent bulb (Exploration 3A) and a fluorescent light (Exploration 4D). How do the theories (blackbody model and quantum model) that explain the spectra of these everyday objects help you explain the spectrum of a star?

Session 5   What is in a star?

Culminating Project

What do we know about stars from stellar spectra?

Throughout this module we have learned how to extract information about the temperature and composition of a star from its spectrum. This culminating project asks you to apply the knowledge and skills you have learned to the analysis of a group of different stars.

Your instructor will provide you with 9 stellar spectra (numbered 1-9). You will receive *two* copies of each. One copy will have peaks labeled by arrows in the range 390-670 nm. The other copy will be the same stellar spectrum, but plotted over a larger range of wavelengths.

You will work both individually and in groups of 3 for this project. The first part of the project requires you to work individually, analyzing the spectra of several stars. Your group will then come together to share individual results and test your understanding of the information contained in stellar spectra.

INDIVIDUAL STELLAR SPECTRAL ANALYSIS

The first part of this project is an individual effort. Each person in the group should take 3 stellar spectra to analyze fully. Attach your name to the 3 spectra you analyze on your own, and provide your individual answers to the questions below. You will be asked to hand in your individual spectra and answers along with your answers to the group questions. Each person in the group should select one of the following sets of stars:

- Set A: 1, 2, 3
- Set B: 4, 5, 6
- Set C: 7, 8, 9

NOTE: To answer questions 1 and 2 below, use the spectrum *without* arrows. In questions 3 and 4, focus your attention on the spectra *with* arrows.

For each of your 3 stars, answer the following questions:

1. Estimate the temperature of your star using its spectrum. Explain your reasoning and show a calculation with units to support your answer. Also, briefly indicate how sure (or not) you are of this temperature estimate and why.

2. How might you determine the expected color of your star? State your reasoning. **Note:** You do not actually have to determine star color, only explain how you would do it.

3. Identify which elements are in your star. Briefly explain your reasoning process and support your identifications with data (a table form is ideal). Label your spectra with measured wavelengths, too, for easy visual comparison when you join your group.

4. Write the electron configuration of all the *atoms* and *ions* in your stars. You do not need to repeat this for elements that are in more than one star.

GROUP QUESTIONS AND ANALYSIS

Schedule a group meeting and come prepared with your individual results. You can only succeed as a group if everyone has analyzed his or her individual stars. This part of the project will be written up as a group response and turned in with all members' names on it. Your group may select a scribe(s) to record responses to the group questions below; however, note the name of the scribe on each question.

5. Share individual results for your 3 stars with your team members. You will have to teach each other about your individual set of 3 stars. Check each others' work for logic, explanations, and correctness.

    NOTE: You may modify your individual answers if you change your mind about something based on group discussion, but note both your original answer and how you changed it.

6. As a group, rank all 9 stars in order of increasing temperature. Explain your reasoning. Is there any uncertainty in this ranking, and if so, why? (You need to be confident of your ranking before going on to the next two questions.)

7. Consider the ions in stars:

    a. Describe the overall trend in the prominence of Ca^+ and He^+ ions for the set of 9 stars. Do you notice these ions in all 9 spectra?

    b. Use your understanding of the electron configurations and relative sizes of Ca and He atoms to explain the difference in ionization energy between Ca and He. Then, use this information to explain the trends you noted in part 7a for Ca^+ and He^+. Your answer should also include consideration of star temperature (see page 46).

    c. Do you expect to see absorption lines due to H^+ in a stellar spectrum? Explain why or why not. (Hint: What effect does absorption of energy have on electrons within an atom or ion?)

8. Consider hydrogen in stars:

    a. Catalogue in a table all the H absorption lines in the group of 9 stars and assign them to specific transitions in H with the notation n_i --> n_f. (You do not need to repeat this for H lines that are in more than one star.) Note the equation that you used to support your assignments.

    b. Draw an energy level diagram of hydrogen that shows the transitions (using arrows) occurring in H atoms in stars. Finally, draw a cartoon model of a star that explains how these spectral lines are formed.

    c. Look at the prominence or strength of the H lines in the star spectra. Describe the trend you see as you move from cool to hot stars.

    d. Make a proposal for why some of the 9 stellar spectra have weak H Balmer absorption lines while others have strong Balmer lines. Consider star temperature, our model for a star, and the atomic processes that create the lines in the spectrum. Assume that hotter stars provide more thermal energy (page 46) to bump electrons in atoms away from the nucleus, and that there can be enough thermal energy in a star to ionize atoms. (Hint: What do the Balmer absorption lines of hydrogen have in common, according to Problem 10 on page 42? You may want to use your reasoning from Problem 7c above.)